Praise

"It is time that a boo[...] quate state of science on the verge of collapse. This attack does not aim to knockout with one blow bad practice of contemporary science; it rather circles the issue and delivers body blows until the job is done. [...] It starts with the epidemic of retractions and failure across science. It debunks evidence-based science. It goes on to analyze the history of how science got into this mess. Gratuitous mathematization is exposed and stands naked. And finally attitudes to doubt and certainty are laid bare, as contemporary science loses every round, left on the verge of collapse."

Professor Emeritus Timothy Allen
University of Wisconsin–Madison

"A major contribution, this book approaches the current crisis of scientific practices with deep insights on the entanglement of science, policy and ethics."

Professor Bernadette Bensaude-Vincent
Université Paris 1, Panthéon Sorbonne

"This book is about complex issues in science and governance relationships that need clarification. It is a fundamental contribution that should interest scientists, policy makers, practitioners and theoreticians involved in evidence-based decision making. In particular, it deals with the interface between research and policymaking, investigating some important areas where more research and discussions are needed. The book poses key questions and provides some answers. As such, the book is relevant to researchers and policy makers alike."

Professor Ron S. Kenett
University of Turin (Italy)
Founder & CEO, the KPA Group (Israel)

"The 'crisis' in science is not emergent: it has been brewing at least since WWII, if not since the scientific revolution. And these authors have been pointing to it for nearly half a century, noting that the low quality of so much scientific output results from a 'structural contradiction' in how science is conducted, funded and governed, socially, economically and morally. […] Read it and weep. And commit to doing better."

Professor Philip B. Stark, Associate Dean
Division of Mathematical and Physical Science
University of California Berkeley

"It is too easy when we talk about science to get nostalgic, to imagine a republic of independent, moral agents working for what Francis Bacon called 'the relief of man's estate'. We need books like this to remind us that 21st century technoscience is big business. […] The litany of controversies, corporate distortions, ethical missteps, retractions, impact factors, league tables and other vices is lengthening. As economies become 'knowledge economies' and governments discover new forms of technocracy, we mustn't pretend that 'pure' science is not politicised and marketised. This book offers an uncomfortable but vital diagnosis of the trouble with science."

Professor Jack Stilgoe, Senior Lecturer
University College London

"With environmental and social imperatives growing apace and globally structured interests increasingly obscuring the picture, the authors provide an authoritative, inspiring and highly readable vision. In many rigorous but practical ways, they show how science and democracy can be mutually reinforcing—and work more effectively together towards vital solutions."

Professor Andy Stirling
SPRU and STEPS Centre, University of Sussex

The Rightful Place of Science:
Science on the Verge

Contributors

Alice Benessia
Silvio Funtowicz
Mario Giampietro
Ângela Guimarães Pereira
Jerome Ravetz
Andrea Saltelli
Roger Strand
Jeroen P. van der Sluijs

Consortium for Science, Policy & Outcomes
Tempe, AZ and Washington, DC

THE RIGHTFUL PLACE OF SCIENCE:
Science on the Verge

The Rightful Place of Science series explores the complex interactions among science, technology, politics, and the human condition.

For information on the Rightful Place of Science series,
write to: Consortium for Science, Policy & Outcomes
PO Box 875603, Tempe, AZ 85287-5603
Or visit: http://www.cspo.org

Model citation for this volume:

Benessia, A., Funtowicz, S., Giampietro, M., Guimarães Pereira, Â., Ravetz, J., Saltelli, A., Strand, R., and van der Sluijs, J. P. 2016. *The Rightful Place of Science: Science on the Verge*. Tempe, AZ: Consortium for Science, Policy & Outcomes.

Other volumes in this series:

Pielke, Jr., R. 2014. *The Rightful Place of Science: Disasters and Climate Change*. Tempe, AZ: Consortium for Science, Policy & Outcomes.

Zurilnik, M. L., *et al.* 2015. *The Rightful Place of Science: Creative Nonfiction*. Tempe, AZ: Consortium for Science, Policy & Outcomes.

ISBN: 0692596380

ISBN-13: 978-0692596388

LCCN: 2016932268

FIRST EDITION, MARCH 2016

CONTENTS

PREFACE

This book is the result of a long-standing collaboration between the creators of the theory of post-normal science, Silvio Funtowicz and Jerome Ravetz, and a group of close collaborators in Italy, Norway and the USA. What we have in common is our interest in theoretical, critical research and interdisciplinary, practice-oriented experience in which we "get our hands dirty" to improve actual practice in science and governance. This is at least partly the outcome of long experience in working closely with and within governance institutions at the national and international level, in particular the institutions of the European Union, whose staff we have served by providing training in how to use evidence for policy and by drafting guidelines and handbooks on the subject.

We would like to mention the four research environments that have been instrumental in the creation and development of this growing research community: (1) the Joint Research Centre of the European Commission in Ispra, Italy, where Ângela Guimarães Pereira is currently based, Andrea Saltelli and Silvio Funtowicz worked for many years, and Jerome Ravetz and Alice Benessia have been frequent guests; (2) the Centre for the Study of the Sciences and the Humanities at the University of Bergen, Norway, which has become a stronghold of post-normal science in Europe and the home institution of Silvio Funtowicz, Matthias Kaiser, Jeroen van der Sluijs and Roger Strand; (3) the Institute of Environmental Science and Technology at the Autonomous University of Barcelona and notably Mario Giampietro's research group on Multi-Scale Integrated Analysis of Societal and Ecosystem Metabolism; and last but not least, (4) the Consortium for Science, Policy & Outcomes coordinated from Arizona State University and co-directed by Dan Sarewitz. In 2015, we established the European Centre for Governance in Complexity, co-directed by Andrea Saltelli, Roger Strand and Mario Giampietro. While still in its infancy, this centre should, with time, become one of the locomotive forces in the theoretical and practical developments which this book exemplifies and advocates.

As is clear in the present work, we try to call attention to the existence of some worrying fault-lines in the present use of science for governance, which link dramatically to the crisis in the governance of science itself—a connection ignored by most authors and commentators. Ethical, epistemological, methodological and even metaphysical dimensions of the crises are identified for the consideration of our readers. Some modest elements of possible solutions are also put forward, especially in relation to what needs to be *un*learned in order to begin the process of reconstruction, and which craft skills need to be relearned, sustained or enhanced.

Acknowledgment

The authors are indebted to Sarah Moore for checking several versions of the present volume, providing critical insight and constructive suggestions, and keeping a keen eye on the coherence among the different parts of the text. Any remaining flaws are the responsibility of the authors.

A note on the text

This volume of *The Rightful Place of Science* follows the University of Cambridge spelling and punctuation style.

FOREWORD

Daniel Sarewitz

Women on the Verge of a Nervous Breakdown was a funny title for Pedro Almodóvar's funny movie about how men drive women nuts. *Science on the Verge* is a funny title, too, but this book, which examines the unfolding crisis in science today, is serious. And indeed, the worrisome, in some ways even terrifying state of affairs in science revealed here, demands the sober, rigorous and intellectually compelling treatment that you are about to read.

And yet... science's problems seem also to verge naturally toward an encounter with satire. If science is the great social enterprise that separates the modern, rational human from our primitive, superstition-laden forebears, how could it have so lost its grip on reality?

The satirical potential that such a question raises has not gone entirely unnoticed, although I can think of only one seriously good satire about the scientific endeavour itself: Jonathan Swift's *Gulliver's Travels* — now nearly 300 years old. Best known for tales in which its itinerant hero is little among big people, big among little people, and a brutish human among apparently civilized horses, *Gulliver's Travels* also recounts the visit of its ingenuous and reliable narrator to the floating (in air) island of Laputa, a kingdom ruled by mathematicians, that most logical and disciplined species of intellect. In Laputa, the nation's indolent leaders are not fanned by servants with palm fronds (as would befit your standard Pharaoh or

iii

Sultan). Rather, servants must continually flap the ears and mouths of their masters with "a blown Bladder fastned like a Flail to the End of a short Stick" in order to get their attention. Otherwise their minds "are so taken up with intense Speculations, that they neither can speak, or attend to the Discourses of others".

Gulliver visits the Academy of Lagado, the kingdom's scientific institute, and describes many projects being pursued by the kingdom's visionary researchers. Here, in an 18th-century nuclear fusion lab, one scientist has spent eight years trying to extract sunbeams from cucumbers. He is confident that, with an additional eight years of work, his project will achieve its goal of storing the extracted energy in "Vials hermetically sealed", so that they can, when needed, be "let out to warm the Air in raw inclement Summers." Meanwhile, the Academy's behavioural economists debate the best way to raise taxes "without grieving the subject. The first affirmed, the justest Method would be to lay a certain Tax upon Vices and Folly […] The second was of an Opinion directly contrary; to tax those Qualities of Body and Mind for which Men chiefly value themselves". Even 'big data' is very much on the agenda, as one especially ambitious professor strives to increase the productivity of scientific research with a huge machine that randomly combines "all the Words of Their Language in their several Moods, Tenses and Declensions", and through this device "give the World a compleat Body of all Arts and Sciences", an effort that would be greatly expedited if only "the Publick would raise a Fund for making and employing five Hundred" such machines.

And what of the world portrayed in *Science on the Verge*? In this book you will read about a scientific enterprise that is growing in productivity and influence even though the majority of publications in many scientific fields may be wrong. You'll see how scientists reduce complex, unpredictable problems to much simpler, manageable models by leaving out important factors, which allows the scientists to come up with neat solutions—often to the wrong problems. You'll learn how doing this sort of science often makes our knowledge of the world more uncertain and unpredictable, not less, and how instead of leading to 'evidence-based policy' we end up with 'policy-based evidence.' You'll find out why precise quantitative estimates of some of the impacts of climate change are so uncertain as to be meaningless. (How, for example, can we quantify to a tenth of a percent the proportion of species that will go extinct from climate change if we don't even know the number of species that exist now?) And you'll find out how economic analyses based on flawed computer coding served the interests of both economists and policy makers—and as a result caused long-term damage to national economies. You'll discover how, in a human world that is growing ever more complex, our approaches to governing science and technology are turning decisions and action over to computer algorithms and technological systems. We transfer our agency to machines in the name of efficiency and predictability, but the entirely paradoxical consequence is that the human capacity to adapt to uncertainty and unpredictability may actually be diminishing.

It's a world that might well have been imagined by a modern-day Swift—only it's our world, today. At its heart is a failure to recognize that the use of science in guiding human affairs is *always* a political act. It's not that we shouldn't do our very best to understand our world as a basis for acting wisely in it. It's that such understanding has its limits as matters of both science and subjective sensibility. All complex systems must be simplified by scientists to render them analytically tractable. All choices about how a society should best address its many challenges must be guided by the norms and values of stakeholders, by trade-offs among those with conflicting goals, and by hedges against inevitable uncertainties. If the second condition — the necessity of subjective choice — is made subservient to the first — the limits of science — then science runs the risk of being corrupted. This happens because its practitioners, advocates and institutions do not resist the temptation of overstating science's claims to both certainty and legitimacy. The risk for society, in turn, comes from pushing the political into the black box of the technical, thus making it invisible to democratic actors. As explained by the political theorist Yaron Ezrahi in his 1990 book *The Descent of Icarus*, "The uses of science and technology to 'depoliticise' action have been among the most potent *political* strategies in the modern state. The authority of this strategy has been sustained by the illusion that social and political problems like scientific problems are inherently solvable" (51).

If science is failing, then, surely a good part of the explanation is that, in turning many complex social challenges over to scientists to find 'solutions', politicians and citizens alike are demanding more from science than it can deliver. Swift himself feared the consequences of substituting scientific rationality for human judgment. Three years after writing *Gulliver*, he explored the problem of scientific rationality and social choice in his famous essay "A Modest Proposal". Here, in a brutal satire of evidence-based policy, he demonstrated in dispassionate, rational, quantified scientific terms that eating poor children would be economically and socially beneficial — a logically elegant solution to poverty arising from England's oppressive policies toward Ireland.

If we have come less far than we might wish from Swift's view of science and politics, the authors of *Science on the Verge* lay out the regimen necessary for avoiding nervous breakdown. Above all is the importance of recognizing that (as you'll read in Chapter 1) "the problems in science will not be fixed by better training in statistics, better alignment of incentives with objectives, better regulation of copyright" and so on. The scientific community continues to understand itself as a self-correcting, autonomous enterprise, but the knowledge it creates is no longer containable within laboratories, technical publications and patents. It has now become central to many political debates, and can be wielded by everyday citizens during activities as mundane as visiting a doctor, buying food or arguing with one's neighbour. Scientists can no longer maintain authority by insisting that they should be left alone to fix their problems. Recall what happened when the Catholic Church tried this approach after Gutenberg had loosened its hold on truth.

Women on the Verge of a Nervous Breakdown made the case for the essential and redemptive strength of women in a male-dominated culture. *Science on the Verge* is no less sympathetic to its subject. Many modern institutions and practices have been designed in the expectation that science was a truth-telling machine that could help overcome fundamental conditions of uncertainty and disagreement. The painful lesson of recent decades, however, is that real science will never construct a single, coherent, shared picture of the complex challenges of our world — and that the quest to do so instead promotes corruption of the scientific enterprise, and uncertainty and suspicion among decision makers and engaged citizens (exemplified in debates over GMOs or nuclear energy). At its best, however, science can provide a multiplicity of insights that may help democratic societies explore options for navigating the challenges that they face. Put somewhat differently, *Science on the Verge* explains to us why science's gifts must be understood as actually emerging from science's limits — much as grace is born from human fallibility.

1

WHO WILL SOLVE THE CRISIS IN SCIENCE?

Andrea Saltelli, Jerome Ravetz and Silvio Funtowicz

Endangered science

The integrity of science has been the subject of increasing concern, among scientists and the media, over the past decade. Although science is still among the most trusted of public institutions, the crisis of quality within science is now threatening to erode that trust.

Attempting to explain "Why most published research findings are false", Ioannidis (2005) expresses "increasing concern that in modern research, false findings may be the majority or even the vast majority of published research claims". The same author has created a Meta-Research Innovation Centre (METRICS) at Stanford to combat "bad science" (*Economist*, 2014). In a later paper Ioannidis (2014) estimated that as much as 85% of research funding is wasted as a result of shoddy science — a serious claim for an enterprise that absorbs sizeable portions of public expenditure. According to *The Lancet* (2015) — which in 2014 ran a series on "Research: increasing value, reducing waste" — an estimated $200 billion was wasted on science in the U.S. in 2010 alone.

"Unreliability in scientific literature" and "systematic bias in research" have been denounced by Boyd (2013); in the field of organic chemistry Sanderson notes that "Laboratory experiments cannot be trusted without verification" (2013); Begley (2013: 433-434) decries "Suspected work [...in] the majority of preclinical cancer papers in top tier journals". In an earlier paper Begley and Ellis (2012) note that a team of researchers working to reproduce 'landmark studies' in haematology and oncology were able to reproduce only 11% of the original findings.

Natural and medical sciences are not the only domains concerned. Nobel laureate Daniel Kahneman was among the first to sound the alarm in the field of behavioural sciences, warning that he saw "a train wreck looming" (quoted in Yong, 2012). Another Nobel laureate, Joseph Stiglitz, described the mathematical work associated with the financial instruments at the origin of the present recession as having been corrupted by "perverse incentives and flawed models — accelerated by a race to the bottom" (2010: 92).

Fraud and misconduct may be only part of the greater problem of integrity in science, but they are the most visible signs of a crisis. The retraction by the journal *Science* of findings according to which political activists could convince conservative voters to change their minds on same-sex marriage in brief face-to-face conversations prompted the *New York Times* (2015) to headline with "Scientists who cheat", and *Nature* (2015), with "Misplaced faith". Citing the finding that nine in ten British people would trust scientists to follow the rules (according to an Ipsos MORI poll), the *Nature* editorial asked pointedly, "How many scientists would say the same?"

Various initiatives have been launched to record misconduct and track retractions (for example, "Retraction Watch"[1]), while problems of reproducibility have been addressed by, for example, the "Reproducibility Initiative"[2] (*Nature Biotechnology*, 2012). However, the process of checking which results are reproducible and which are not is arduous, as reluctance to share data hampers the reproducibility effort (Van Noorden, 2015).

Four World Conferences on Research Integrity have been held between 2007 and 2015 (see *The Lancet*, 2015, for a discussion); the issues are debated in think tanks (Horton, 2015); and as recently as October 2013 the weekly news magazine *The Economist* dedicated its cover page to the subject, with an editorial exploring "How science goes wrong":

> *Science still commands enormous – if sometimes bemused – respect. But its privileged status is founded on the capacity to be right most of the time and to correct its mistakes when it gets things wrong. [...] The false trails laid down by shoddy research are an unforgivable barrier to understanding.* (Economist, 2013: 11)

The crisis in reproducibility must imply a dysfunction in science's own quality control mechanisms. In 2015 the publisher Springer and the Université Joseph Fourier released 'SciDetect', software designed to detect fake scientific papers (Springer, 2015): "The open source software [SciDetect] discovers text that has been generated with the SCIgen computer program and other fake-paper generators like Mathgen and Physgen". In China, "A *Science* investigation has uncovered a smorgasbord of questionable practices including paying for author's slots on papers written by other scientists and buying papers from online brokers" (Hvistendahl, 2013).

---

[1] See http://retractionwatch.wordpress.com

[2] See http://www.reproducibilityinitiative.org

A superficial reading of the evidence could yield the impression that the main cause of the reproducibility crisis is the deficient statistical competence of scientists themselves. This is the line taken by *The Economist* (2013) in the issue quoted above and which is mostly based on a reading of Ioannidis (2005). According to this reading, the theory and practice of P-values, false positives and false negatives should be taught more rigorously to practicing scientists. However, for some commentators, "P-values are just the tip of the iceberg" (Leek and Peng, 2015), and more fundamental problems pervade the whole data-based evidential chain. In addition to statistical skills, issues surrounding incentives are also prominent in the literature quoted above (Ioannidis, 2005; Stiglitz, 2010).

While at the beginning of the integrity movement (marked by the first World Conference in 2007) the main concerns were misconduct and fraud, the proliferating number of individual cases was soon recognized to be symptomatic of a more general malaise. The focus then shifted to "strengthening research integrity and responsible conduct of research worldwide" (*Lancet*, 2015).

Solutions to the crisis from within the scientific community

In this section we look at 'remedies from within' — that is, recipes that have been put forward from the affected community itself (*e.g.* scientific editors, scientists and research institutions) to tackle the crisis. There has been no lack of effort to find remedies for the perceived problems in science. Worthy and ingenious suggestions for repair and reform of the quality assurance system are regularly produced and widely discussed. A notable example is the San Francisco Declaration on Research Assessment, a manifesto drafted by a group of editors and publishers of scholarly journals who convened during the Annual Meeting of the American Society for Cell Biology (ASCB) in San Francisco in December 2012[3]. As of June 2015 the declaration has been signed by 12,377 individuals and 572 organizations. For *The Lancet* (2015) the most relevant recommendation in the Declaration is, *verbatim*, "Do not use journal-based metrics, such as Journal Impact Factor, as a surrogate measure of the quality of individual research articles to assess an individual scientist's contributions, or in hiring, promotion, or funding decisions." The link between the use of better and more diverse (including more qualitative) metrics to appraise researchers' work on one hand, and transparency in the rules for grant allocation, hiring and promotion on the other, is central to the recommendations contained in the Declaration. The recommendations are arranged into five groups: general, for institutions, for publishers, for organizations that supply metrics and for researchers. Several recommendations are common to two or more groups.

While some of the recommendations in the San Francisco Declaration would be straightforward to implement (for example, enlarging the list of the metrics commonly used to appraise journals, to include "5-year impact factor, EigenFactor [...], SCImago [...], h-index, editorial and publication times"), others would be more difficult and potentially contentious, as they concern the measurement of the impact of research on practice and policy:

> *For the purposes of research assessment, consider the value and impact of all research outputs (including datasets and software) in addition to research publications, and consider a broad range of impact measures including qualitative in-*

---

[3] See http://am.ascb.org/dora/

dicators of research impact, such as influence on policy and practice.

Examples of recommendations for publishers are:

Whether a journal is open-access or subscription-based, remove all re-use limitations on reference lists in research articles and make them available under the Creative Commons Public Domain Dedication. [...] Remove or reduce the constraints on the number of references in research articles, and, where appropriate, mandate the citation of primary literature in favor of reviews in order to give credit to the group(s) who first reported a finding.

As mentioned, there are parallels between the recommendations addressed to the various groups, so those specific to researchers include suggestions on how to behave when involved in committees making decisions about funding, hiring, tenure or promotion. Also recommended are: the citing of primary literature; the use of a broad range of criteria to evaluate the research of peers; and the questioning of research assessment practices that rely inappropriately on 'Journal Impact Factors'.

A checklist of good practice from the perspective of a researcher is offered by Ioannidis (2014) in his paper, "How to Make More Published Research True". His recommendations start with scientific practices:

To make more published research true, practices [should] include the adoption of large-scale collaborative research; replication culture; registration; sharing; reproducibility practices; better statistical methods; standardization of definitions and analyses; more appropriate (usually more stringent) statistical thresholds; and improvement in study design standards, peer review, reporting and dissemination of research, and training of the scientific workforce.

Another of his suggestions is to apply a scientific method (specifically counterfactual verification) to what works in science itself:

Selection of interventions to improve research practices requires rigorous examination and experimental testing whenever feasible.

Two further recommendations from Ioannidis target the system of incentives:

Optimal interventions need to understand and harness the motives of various stakeholders who operate in scientific research […].

Modifications need to be made in the reward system for science, affecting the exchange rates for currencies (e.g., publications and grants) and purchased academic goods (e.g., promotion and other academic or administrative power) and introducing currencies that are better aligned with translatable and reproducible research.

Most of the recommendations listed above by way of example are unproblematic: who would argue against more transparency, less reliance on metrics that can be manipulated, or a less stringent licensing system to protect authorship? It also seems logical that the system of 'currencies' regulating research careers should be in alignment with the type of science one wishes to foster. At the same time, we suggest that a tendency to focus on the issues surrounding incentives as if these were core of the problem may lead us to overlook deeper and more fundamental factors; if the prevalence of low-quality research were in fact a manifestation of a state of corruption in science, the cure would not be only a matter of improved arrangements for collaboration, inspection or regulation (Ravetz, 1971: 407).

This use of the word 'corruption' above does *not* imply that most, or even many, scientists consciously purvey false or faulty results for some undeserved reward. Rather, it refers to a situation in which, because of changing social arrangements and ethical frameworks, and the consequent discrepancy between the image and the reality of scientific life, it becomes increasingly difficult for scientists to do the good work to which they would normally aspire. In every socially organized activity there are pressures on individuals to pursue short-term personal gains (or to seek protection) at the expense of higher goals. The quality assurance system in science has the function of protecting its practitioners from those corrupting pressures, by implementing testing routines which are supported by incentives and sanctions, backed up by informal peer pressure and validated by the leadership of exemplary individuals (Ravetz, 1971: 22-23). This is clearly a complex social system whose effective functioning cannot be guaranteed by administrative means alone. An apparent confirmation of the systematic character of the corruption problem has been provided by the leaked dossier on fraud in British science reported by the *Times Higher Education* (Matthews, 2015).

The fact that it appears to be extremely difficult to find effective solutions (Horton, 2015) suggests that the problem of quality in science may have its roots in a "structural contradiction" in the very system of production of scientific knowledge (Ravetz, 2011).

Rethinking the problem

In this section we suggest that the crisis in science has not yet been accurately described or diagnosed and that real insight into the situation will require a deeper analysis of its causes.

The root of the crisis could well lie in the very success of science. In his 1963 work *Little Science, Big Science*, de Solla Price anticipated that science would reach saturation (and in the worst case senility) as a result of its own rapid exponential growth, in terms of numbers of researchers, publications, volume of outlays, *etc.* (1963: 1-32). De Solla Price was a perceptive witness of the transformation of science. Though the distinction between little and big science was inspired by an earlier work of Weinberg (1961), De Solla Price had a clear understanding of how the social organization of science and quality control systems would have to change under the new conditions of post-war industrialized science.

De Solla Price is considered today the father of scientometrics, although his forewarnings of the impossibility of the endless growth of science and the implicit dangers in the change in the status of researchers received relatively little attention at the time. Without a doubt, a major cause of the present difficulties is the sheer scale of big science. As personal contact among researchers in the same field has become impossible, scientific communities are less cohesive than before. Big science also engenders the need for 'objective' mass metrics of quality, which are inevitably imperfect and often perverse and corruptible. These effects are compounded by new economic and commercial pressures, in a social and cultural context in which the idealism that motivated 'little science' is no longer compelling.

Disruptive qualitative changes in the conduct of research have been identified by historian and philosopher of economic thought Philip Mirowski as the effects of the 'commoditization' of science. Mirowski offers his detailed aetiology of the predicaments of science in his 2011 book *Science-Mart: Privatizing American Science*. He makes reference in the title to the aggressive giant supermarket chain Walmart, which epitomizes a culture and ideology that, in Mirowski's diagnosis, represent a large part of the problem. According to Mirowski, since the 1980s neoliberal ideologies have succeeded in establishing the principle that the market is the best answer to the question of how best to allocate resources — including in the scientific domain. State funding of research has accordingly decreased and major corporations have closed down their in-house laboratories and begun to commission research, first to universities and then to contract research organizations (CROs), which operate under significant budget and time pressures.

Mirowski's argument is that the quality of science suffers under conditions of commoditization and that this is now undermining the capacity of science to produce innovation. The effects are most obvious in the intrusion of property rights into the materials and outputs of research, to the extent that scientists are often mired in application processes to obtain the requisite permissions, while administrators are preoccupied with processing such applications. In the terms of Mirowski's analysis, the perverse or distorted system of incentives described above is a collateral effect of the prevailing neoliberal ideology. Such a regime is favourable to the 'entrepreneurial scientist', whose career is defined by successful grants enabled by adequate projects, rather than the other way around as in the days of 'little science' (Ravetz, 1971: 46). The use of the term 'currency' by Ioannidis, cited above, suggests that the commoditization of science is so thoroughgoing that even suggested correctives are expressed in the discourse of a neoliberal paradigm. As Mirowski notes, even the quantitative indicators of quality are controlled by the private sector, meaning that we cannot really know how well, or how ill, science is doing. This problem is reaffirmed in a recent discussion in *Nature* (Wilsdon, 2015).

It is evident that the system of incentives applying to medical researchers in a CRO will differ from the system governing researchers working in a national laboratory. Nevertheless, it could be argued that neoliberalism has

simply accelerated a process which was already under-
way and which had long been foreseen by scholars of
science and technology. Indeed, Mirowski's historical
account could be interpreted as the institution of American science struggling to maintain its high rate of
growth, first by cashing in on the prestige it earned from
World War II, and then by finding industrial sponsor-
ship. But as De Solla Price had anticipated, the limits of
federal largesse were eventually hit, at some moment in
the 1990s. Since then the constriction in funding has intensified the crisis in the conditions of work for researchers, and (more worryingly) has contributed to an
ageing of the population of working scientists. As Colin
Macilwain says (2015: 137), "it is increasingly older people, who know how to work the system, who get funding: people under 40 are finding it harder and harder to
get their foot on the ladder". A life in science has
evolved from being a vocation, to being a career, and
finally, in the 21st century, to being an insecure job.

In his book *Scientific Knowledge and its Social Problems*
(1971), Ravetz foresaw that serious trouble for science's
quality assurance mechanisms and for morale in the scientific community would follow from the debased ethos
of industrialized science:

> *[...] with the industrialization of science, certain changes
> have occurred which weaken the operation of the tradition-
> al mechanism of quality control and direction at the high-
> est level. [...]The problem of quality control in science is
> thus at the centre of the social problems of the industrial-
> ized science of the present period. If it fails to resolve this
> problem [...] then the immediate consequences for morale
> and recruitment will be serious; and those for the survival
> of science itself, grave. (1971: 22)*

> *Two separate factors are necessary for the achievement of
> worthwhile scientific results: a community of scholars with
> a shared knowledge of the standards of quality appropriate*

11

for their work and a shared commitment to enforce those standards by the informal sanctions the community possesses; and individuals whose personal integrity sets standards at least as high as those required by their community. If either of these conditions is lacking – if there is a field which is either too disorganized or too demoralized to enforce the appropriate standards, or a group of scientists nominally within the field who are content to publish substandard work in substandard journals – then bad work will be produced. This is but one of the ways in which 'morale' is an important component of scientific activity; and any view of science which fails to recognize the special conditions necessary for the maintenance of morale in science is bound to make disastrous blunders in the planning of science. (1971: 22-23)

The same section of the book discusses the technocratic view of science to argue that the assimilation of the production of scientific results to the production of material goods can be dangerous, and indeed destructive of science itself. One of the main thrusts of this early work is in fact the illustration of the paradox that the successful production of objective scientific knowledge depends critically on the individual moral commitment of scientists themselves. The question, "*Quis custodiet ipsos custodes?*" (Who guards the guardians?), haunts the quality assurance system in science, the vulnerability of which is exacerbated by its reliance on iterative, informal and ultimately judgemental procedures.

It should be stressed that the "shared commitment" quoted in Ravetz (above) is a far cry from the "better incentives" advocated by would-be reformers within the scientific establishment (*e.g.* in the San Francisco Declaration). While material rewards certainly have their place in the maintenance of morale in a professional or fiduciary activity, it would require social engineering of a high order to make them the sole wellspring of ethical commitment.

At the heart of Ravetz's reflections is the recognition of the need for good morale in science:

The need for good morale is never mentioned in general discussions of science directed to a lay audience; and this is evidence that hitherto its presence could be taken for granted. For doing good scientific work is strenuous and demanding, and the quality of the work done in any field of science is dependent, to a great extent, on the integrity and commitment of the community of scientists involved. (1971: 58)

Ravetz goes on to outline the hypothetical situation that the peer review mechanism is intended to avert:

If there were not a test of each paper before its acceptance by a journal, then every intending user would be forced to examine it at length before investing any of his resources in work which relied on it. Under such circumstances, the co-operative work of science as we know it could not take place. (1971: 176)

In effect, this highly undesirable situation has to some extent materialized: some chemists have felt themselves obliged to replicate organic syntheses since they cannot trust published results (Sanderson, 2013).

Can science's predicaments be resolved, and how? Ravetz's understanding of the matter is as follows:

No formal system of imposed penalties and rewards will guarantee the maintenance of quality, for the tasks of scientific inquiry are generally too subtle to be so crudely assessed; nor will the advantages to an individual of a good reputation of his group be sufficient to induce a self-interested individual to make sacrifices to maintain it. Only the identification with his colleagues, and the pride in his work, both requiring good morale, will ensure good work. (1971: 407)

> *The conditions of industrialized science present [leading scientists] with problems and temptations for which their inherited 'scientific ethic' is totally inadequate. (1971: 408)*

Our conclusion thus far is that the problems in science will not be fixed by better training in statistics, better alignment of incentives with objectives, better regulation of copyright or the elimination of impact factors, although these and other measures discussed in Section 2 all have their uses. We are not dealing with isolated crises in reproducibility, peer review mechanisms or hiring practices; rather, we are facing what we could call a generalized crisis in the epistemic governance of science.

A rather severe judgement—taking the epistemic governance crisis to an extreme conclusion—is offered by Millgram (2015: 21-53). To Millgram, the success of the Enlightenment is the root cause of its ultimate failure—that is, the failure of man's free will and independent judgement to triumph over irrational principles of authority, religion and superstition. What the Enlightenment has in fact generated is a society of "serial hyperspecializers" (26), a world in which all knowledge and products are the result of some form of extremely specialized expertise, and in which expertise is itself highly circumscribed, since experts depend in turn on other experts whose knowledge claims and styles of argumentation cannot be exported from one discipline to the next. Experts thus become "logical aliens" (32) to one another and humans become incapable of forming judgements. The author describes this as "the great endarkenment", characterized by "commitments (both decisions and views as to how the facts stand) whose merits no one is in a position to assess—where […] each decision, […] is made on the basis of further decisions, whose merits no one can assess either. Trusting in the outputs of this process is on a par with settling what you are going to do by reading entrails or casting hexagrams" (36-37). To make decisions in this environment, one needs constantly to wrestle quality "from the jaws of entropy". Millgram contends that we simply have no idea of how bad the situation is, likening it to the era of premodern medicine, under which patients endured pain and suffering because "they weren't equipped to assess the theories, inferential practice, and effectiveness of the procedures performed by members of a special-ized professional guild" (37). The world he depicts could well be a portrayal of De Solla Price's vision of the senili-ty of science. Ravetz (2011) provides an analogous his-torical perspective, in terms of the maturing of the structural contradictions of modern science.

What can be done

The main impediment to a possible cure of the pre-sent disease is the belief that the system will straighten itself out: that the scientific community can use its own craft to mend itself. This is implausible because the as-sumptions, structures and practices out of which the crisis arose are not likely to produce its solution unaid-ed. Some help from without is in order.

For *The Lancet* (2015), "The coming together of the three themes — research integrity; research reward sys-tems; and increasing value and reducing waste in re-search — is helpful and has greater potential in effecting change than each on its own."

There is no denying the importance of those themes, but some crucial issues remain unaddressed: primarily that of 'who' will launch and pursue this process and, even more importantly, 'how'. The institutional re-sponse has not yet been adequate. As Colin Macilwain notes in *Nature* (2015), the peer-reviewed paper, "the main yardstick for success or failure in almost all aca-demic research careers", and the peer-reviewed, single-investigator grant, another pillar in the mechanics of the scientific system, have been left untouched (*pace* the San Francisco declaration). In Macilwain's view, both the European Commission and the American National Institutes of Health (NIH) prefer to muddle through with obsolete structures and mechanisms. When top-level medical scientists gather to ponder these problems and their possible systemic solutions, it seems that they are content to be advised by a physicist who recommends standards of statistical significance (1 in 3.5 million) that may be appropriate for high-energy experiments but are hardly applicable elsewhere (Horton, 2015).

If purely technical or instrumental solutions are unlikely to be adequate to solve the crisis, then the intuitions and endeavours of concerned scientists are likely to benefit from the contributions of other voices, including reflective scholars, journalists and members of civil society.

Scientists will certainly play a crucial role in the construction of the future of science. Timothy Gowers's campaign against Elsevier with the slogan "Academic Spring" shows how effective a scientist can be when a new consciousness is achieved (Whitfield, 2012). The "Science in Transition" discussion in the Netherlands[4] is an example of committed scientists taking the initiative. Courageous librarians such as Jeffrey Beall at the University of Colorado, Denver, help in the fight against "predatory publishers" who charge authors for publishing but do not provide any control or peer review. Beall also keeps an eye on other degenerations[5]:

The Misleading metrics list includes companies that "calculate" and publish counterfeit impact factors [...] The Hi-

---

[4] See http://www.scienceintransition.nl/english
[5] See http://scholarlyoa.com/2015/01/02/bealls-list-of-predatory-publishers-2015/

jacked journals list includes journals for which someone has created a counterfeit website, stealing the journal's identity and soliciting articles submissions using the author-pays model (gold open-access).

Beyond these initiatives, we urgently need a philosophy of science in which science's imperfections and vulnerabilities are acknowledged and explored. An early attempt at this was made by Thomas Kuhn (1962), who broke with a venerable philosophical tradition which held that science produced truth or its best approximation. His insights into the contingency of scientific truths were built on by Ravetz (1971), with a focus on quality and on the personal, moral element in the achievement of objective scientific knowledge. A further philosophical development, taking into account complexity, ignorance, abuse of mathematics and corruption in its broad (not personal) sense, is now overdue.

One necessary step on this path will be to review the traditional assumption of the separation between science and society. This has been described elsewhere as the "demarcation model" (Funtowicz, 2006) and as the "Cartesian dream" of infinite perfectibility driven by autonomous science (Guimarães Pereira and Funtowicz, 2015):

For several centuries, the understanding of science has been conditioned by a belief in the separateness of knowledge and society. [...] That simple faith is no longer adequate for its function of maintaining the integrity and vitality of science. (Ravetz, 1971: 405)

There is a rich body of scholarship warning against the delusion of such a separation, from Toulmin's *Return to Reason* (2001) and *Cosmopolis* (1990) to Feyerabend's *Against Method* (1975), Lyotard's *The Postmodern Condition* (1979) and Latour's *We Have Never Been Modern* (1993). Although 'science wars' were fought between the natural and human sciences in the 1980s and 1990s, belief in the integrity of science *vis-à-vis* social and political influence still prevails, not least in the institutions and apparatuses in charge of governing and funding science.

The ideal of a disinterested scientific practice which manages to isolate itself from the messiness of everyday life and politics is, of course, an abstraction. We deal not in pure facts of nature, but in 'hybrid', socially constructed objects (Latour, 1993), and one of the features of the present epistemic governance crisis is that "the more knowledge is produced in hybrid arrangements, the more the protagonists will insist on the integrity, even veracity of their findings" (Grundmann, 2009). Nowhere is such a crisis more evident than in the inappropriate use of mathematical modelling and quantification of the world (Saltelli and Funtowicz, 2014), which undermines the use of such evidence for policy (see Chapter 2, this volume)[6]. While it is clear that the demarcation model aims to protect science from political interference by preventing the potential abuse of science and scientific information for political agendas, the model relies on the ideal of separation between the institutions and individuals that 'do' science and those that 'use' it. Whether or not this ideal is even desirable in abstract terms, it is a chimera. The integrity of science will be better protected by heightened awareness of its vulnerabilities, not by fantasies of isolation.

One of the most hopeful signs in the present crisis is the tacit abandonment of the traditional image of science as a truth-producing machine. For centuries, philosophers and historians preached the inexorable progress of Truth. Students never saw incorrect statements in science other than those resulting from their own stupidity. Instances in which great scientists had been partly or

---

[6] See also https://ec.europa.eu/jrc/en/event/conference/use-quantitative-information

wholly wrong were glossed over. It became nearly inconceivable that research based on numerical data and mathematical methods could be wrong or futile. That has all changed quite recently, as we saw in Section 1 above. There is now a burgeoning literature that shows by example that science is fallible. This is easily demonstrated in the case of statistics, in which the practitioner may legitimately obtain any desired answer by choosing the parameters judiciously (Aschwanden, 2015).

We can therefore already point to some of what needs to be 'unlearned' in the prevailing model of science. Kuhn observed that education in science is "narrow and rigid" and comparable to orthodox theology (1962: 165). There is an 'implicit scientific catechism' that students learn by example but that working scientists must leave behind: chiefly, that every scientific problem has one and only one correct solution, precise to several significant digits; that quantitative data and mathematical techniques produce certainty; and that error in science is the result of stupidity or malevolence. Small wonder that when this Cartesian philosophy, with its reliance on the illusory precision of models and indicators, guides any attempt to understand and manipulate complex systems, it can go so spectacularly wrong.

Education will therefore clearly be an important facet of the necessary reform of science. There are signs that science education may already be changing, under the influence of the new social media. The growth of 'do-it-yourself (DIY) science', showing only a minimum of deference to established science, will eventually influence science education to good effect. When students conceive of a scientific exercise as a 'hack' rather than a 'proof', a new consciousness is being created. Kuhn's gloomy picture of science education may at last be on the way out.

In his quest for social solutions to the anticipated predicaments of science, Ravetz (1971) envisaged a new "critical science" which had considerations of participation and respect for the environment at its heart. These ideals were further developed in subsequent works (Funtowicz and Ravetz, 1990, 1991, 1992, 1993) and led to the concept of "post-normal science" (PNS), which is today relatively well known as an approach to deal with problems at the interface between science and policy. While PNS was designed to apply where facts are uncertain, values in dispute, stakes high and decisions urgent (Funtowicz and Ravetz, 1991, 1992, 1993), with the ecological movement as one of its driving forces, PNS is understood today as a system of epistemic governance of practical applicability to all domains in which science and society interact, *i.e.* by definition, to all settings where science operates, including reflexively to the operation of science itself. In PNS the focus is on participation, legitimacy, transparency and accountability. In the "extended participation model" (Funtowicz, 2006), deliberation (on what to do) is extended across disciplines — in the acknowledgment that each discipline has its own lens — and across communities of experts and stakeholders. In adopting this model, one moves from 'speaking truth to power' towards "working deliberatively within imperfections" (Funtowicz, 2006f).

McQuillan (2014) has recently remarked that the movement known as 'Citizen Science' could seize the opportunity created by the crisis in science. PNS is singled out by McQuillan as a promising framework for the work to be done. The remainder of this section will focus on the possibility that the principles of PNS, foremost that of the participation of extended peer communities, may furnish some elements of a solution to the crisis.

New forms of activity in science are appearing rapidly, due to the interaction of new information technologies and new political currents (Ravetz and Funtowicz, 2015). The introduction of the photocopying machine and then of the internet has dramatically changed the conditions for quality control in science. The gateways and the strictures of the printed page have been eliminated. Anyone can distribute information at an unprecedented speed. The present times could be seen as analogous to the period that followed the invention of the printing press and the publication of Gutenberg's Bible around 1450. Now, as then, the monopoly over the channels of knowledge dissemination is collapsing and new, expanded audiences are being created. Higher levels of literacy, increased use of information and communication technologies and increased awareness of complexity have been identified by the authors of the Stiglitz Commission for the Measurement of Progress (CMEPSP, 2009: 7) as a cause of increased use and production of statistical indicators by an ever wider public. Greater media interest is being given to all forms of scientific consumption, as discussed at the beginning of the chapter in relation to science-centred controversies. We would highlight one other powerful driver of change: increased levels of advocacy, by which we mean the critical activity whereby citizens of varying degrees of scientific literacy take it upon themselves to examine and pronounce on the 'goods' and the 'bads' of science and technology. "Is the internet to science what the Gutenberg press was to the church?" ask Funtowicz and Ravetz (2015).

Daniel Sarewitz (2015) also considers the participation of citizens essential if the application of science to policy is to work. In his paper "Science can't solve it", he argues that questions over crucial issues such as genetically modified (GM) organisms, nuclear power and the efficacy of cancer screening cannot and should not be decided by experts alone, lest the legitimacy of science's role in society be eroded.

Leadership that inspires by precept and example will also be needed. A parallel may be drawn with the introduction of principles of 'total quality assurance' into industrial quality control by W. Edwards Deming in the late 1940s. His concepts were successfully taken up by Japanese car manufacturers before being imported back to the USA several decades later. At the core of Deming's work was the establishment of quality circles, in which the assembly line was transformed into a participatory process. In this way the community of expert quality controllers and evaluators was extended to the entire work force, allowing the latter to rely on and convey its practical and possibly tacit knowledge, experience and commitment. The quality circles also encouraged the practice of 'whistle-blowing', whereby any member of the community, including the assembly line workers, could stop the process if they believed that quality had been compromised (Deming, 1986).

An additional justification for the cultivation of extended peer communities in the quality assurance of science is the importance of trust in the system. Science is at present deeply involved in technology and related policy problems that affect public health and welfare, to the extent that the traditional relations of trust can no longer be taken for granted. If there should be another scandal in high-stakes policy-related science—for example, along the lines of the unjustified assurances of the 'safety of British beef' given by official experts during the BSE ('mad-cow') epidemic—then public trust in scientific probity and science-based advice could be seriously affected. Maintaining the justified trust of the public in science is critically important; however, to do that, it will first be necessary to restore the trust of scientists themselves in their own community and practice.

Attempts to circumvent the need for trust by increasing the level of bureaucratic surveillance or quantified

22

quality control criteria are likely to be counterproductive and ineffective. In effect, extended peer communities are already being created, in increasing numbers. They are called citizen juries, focus groups, consensus conferences, or a variety of other names. They come to life either when authorities cannot see a way forward, or when they know that without a broad base of consensus, no policy can succeed. In their early phases these have been largely top-down initiatives, subject to various pressures; but that scene is also changing.

These communities assess the quality of policy proposals, including their scientific justification, on the basis of the science that they master, combined with their commitment and their knowledge — often local and direct — of the subject matter. They ask the sorts of questions that do not occur to the experts, who necessarily conceive of problems and solutions from within their professional paradigms. The form of the questions can be "what if?", "what about?", and "why this and not that?" The moral force of these extended peer communities, which have the ability to create their own 'extended facts' by questioning the framings of the issue proposed by the incumbent powers, can translate into political influence. The extended facts may include craft wisdom and community knowledge of places and their histories, including their history of interaction with the authorities (Lane *et al.*, 2011), as well as anecdotal evidence, neighbourhood surveys, investigative journalism, and leaked documents (Funtowicz and Ravetz, 2015). Local people can imagine solutions and reformulate problems in ways that the accredited experts do not find 'normal'. Their input is thus not merely that of quality assurance of a scientific evidential base, but of problem-solving in general (Funtowicz and Ravetz, 2015). Sheila Jasanoff (2003, 2007) speaks in this respect of "technologies of humility".

Similarly, the DIY science movement puts scientific matters literally into the hands of interested citizens — the embodiment of Giovan Battista Vico's philosophical programme, conceived in opposition to René Descartes. For Vico, *"verum et factum convertuntur"* — literally 'the true and the made are convertible', thus 'I know it if I can make it'. These movements seem to be driven from within and to operate without need of incentives. Two wings can be identified within the citizen science movement: 'amateur-citizen' and 'activist-citizen', depending on how they interact with established science. The former assist professional scientists in folding proteins or classifying galaxies on online interfaces, while the latter take matters into their own hands when they feel that the existing institutions have failed them (the classic example being Lois Gibbs in the case of the Love Canal toxic waste dump).

The rapidly developing citizen science movement may also provide the elements of a solution to the crisis in quality within science. In principle, a less well organized (and occasionally anarchic) movement should have even greater problems of quality assurance than traditional science, with its established structures of control. But there is evidence of two redeeming features of citizen science. One is high morale and commitment among the citizen-scientists (Newman, 2015); the other is the establishment of appropriate systems of quality assurance (Citizen Science Association, 2012). It is, of course, far too early to say how effective these will be; but at least there are encouraging signs.

This mainly external movement may also eventually lead to the emergence of a 'scientist-citizen' movement within established science itself. Scientist-citizens could engage both with the internal problems of science, such as trust and quality assurance, and with the external challenges relating to the use of science to solve practical (*i.e.* policy) problems. An eloquent argument for the recognition of such scientist-citizens has been made by Jack Stilgoe (2009). The means for this development are already being put into place, with the rapid development of new social media techniques in the communication system of science. With the breaking of the effective monopoly of peer-reviewed journals, participation, transparency and openness are flourishing within the communities of science. This new social reality of free internal criticism is antithetical to the so-called 'deficit model', which assumes that public acceptance of a perfect official science is obstructed only by lay people's deficient scientific literacy (Wynne, 1993). A scientist-citizen approach would commit scientists *qua* citizens to criticism, reflection and action.

Conclusion

This chapter started with the identification of a crisis in the governance of science. We are not referring only to the frequency of retractions, which could be an artefact of a more rigorous editorial policy and increased scrutiny by media, bloggers and practitioners. We also observe the increased frequency of warnings, many mentioned in this chapter, by scientists who are concerned about a future blighted by predatory publishers, fraudulent peer reviewing, and the manipulated, and thus often ineffective, use of scientific evidence for policy. We have offered elements of a possible solution, among which we have privileged avenues external to science's own institutions.

Could citizen science and scientist-citizens together perform the rescue of quality and trust in science? It is much too early to say, and the evidence proposed here in support of the thesis is largely based on anecdote, metaphor and analogy and on the predictions of a book that is now more than forty years old (Ravetz, 1971; see also Chapter 3, this volume, for a circumspect view of citizen science). It is clear that action is being taken on many fronts, both in established science and in its new forms of practice. The restoration of quality in science, and the preservation of trust, will not be accomplished by 'scientific' means alone. We are therefore facing one of the greatest challenges for science of our times.

References

Ashwanden, C., 2015. "Science isn't broken. It's just a hell of a lot harder than we give it credit for". FiveThirtyEight. http://fivethirtyeight.com/features/science-isnt-broken

Begley, C. G., and Ellis, M. E., 2012. "Drug Development: Raise Standards for Preclinical Cancer Research", *Nature*, 483: 531–533.

Begley, C. G., 2013. "Reproducibility: Six red flags for suspect work", *Nature*, 497: 433–434.

Boyd, I., 2013. "A standard for policy-relevant science. Ian Boyd calls for an auditing process to help policy-makers to navigate research bias", *Nature Comment*, 501: 160, 12 September.

Citizen Science Association, 2015. "Introducing the Data and Metadata Working Group", 12 November. http://citizenscienceassociation.org/2015/11/12/introducing-the-data-and-metadata-working-group/

CMEPSP, 2009 (Commission on the Measurement of Economic Performance and Social Progress), http://www.insee.fr/fr/publications-et-services/dossiers_web/stiglitz/doc-commission/RAPPORT_anglais.pdf

De Solla Price, D. J., 1963. *Little science, big science*. New York: Columbia University Press.

Deming, W. E., 1986. *Out of the Crisis*. Cambridge, MA: MIT Press.

Economist, 2013. "How science goes wrong", 19 October: 11.

Economist, 2014. "Combating bad science Metaphysicians. Sloppy researchers beware. A new institute has you in its sights", 15 March.

Feyerabend, P., 2010 (1975). *Against Method*. London: Verso.

Funtowicz, S., 2006. "What is Knowledge Assessment?", in Guimarães Pereira, Â., Guedes Vaz, S. and Tognetti, S. (eds.) *Interfaces between Science and Society*. Sheffield: Greenleaf Publishers.

Funtowicz, S. and Ravetz, J., 1990. *Uncertainty and Quality in Science for Policy*. Dordrecht: Kluwer Academic Publishers.

Funtowicz, S. O. and Ravetz, J. R., 1991. "A New Scientific Methodology for Global Environmental Issues", in Costanza, R. (ed.), *Ecological Economics: The Science and Management of Sustainability*: 137–152. New York: Columbia University Press.

Funtowicz, S. O. and Ravetz, J. R., 1992. "Three types of risk assessment and the emergence of postnormal science", in Krimsky, S. and Golding, D. (eds.), *Social Theories of Risk*: 251–273. Westport, CT: Greenwood.

Funtowicz, S. and Ravetz, J., 1993. "Science for the post-normal age", *Futures*, 31(7): 735-755.

Funtowicz, S. and Ravetz, J. R., 2015. "Peer Review and Quality Control", Wright, J. D., (ed.), *International Encyclopedia of the Social and Behavioral Sciences,* 2nd edition. Oxford: Elsevier.

Grundmann, R., 2009. "The role of expertise in governance processes", *Forest Policy and Economics*, 11: 398-403.

Guimarães Pereira, Â. and Funtowicz, S. (eds.), 2015. *Science, Philosophy and Sustainability: The end of the Cartesian dream*. Routledge series Explorations in Sustainability and Governance. New York: Routledge.

Horton, R., 2015. "Offline: What is medicine's 5 sigma?", *Lancet*, 385: 1380.

Hvistendahl, M., 2013. "China's Publication Bazaar", *Science*, 342: 1035-1039.

Ioannidis, J. P. A., 2005. "Why Most Published Research Findings Are False", *PLoS Medicine*, 2(8): 696-701.

Ioannidis, J. P., 2014. "How to Make More Published Research True", *PLoS medicine*, 11(10), e1001747.

Jasanoff, S., 2003. "Technologies of Humility: Citizen Participation in Governing Science", *Minerva*, 41(3): 223-244.

Jasanoff, S., 2007. "Science & Politics. Technologies of humility", *Nature*, 450: 33.

Kuhn, T. S., 1962, *The Structure of Scientific Revolutions* (1st ed.). University of Chicago Press.

Lancet, 2015. Editorial: "Rewarding true inquiry and diligence in research", 385: 2121.

Lane, S. N., Odoni, N., Landström, C., Whatmore, S. J., Ward, N. and Bradley, S., 2011. "Doing flood risk science differently: an experiment in radical scientific method." *Transactions of the Institute of British Geographers*, 36: 15-36.

Latour, B., 1993. *We Have Never Been Modern*. Cambridge, MA: Harvard University Press. Originally published 1991 as

Nous n'avons jamais été modernes. Paris: Editions La découverte.

Leek, J. T. and Peng, R. D., 2015. "P values are just the tip of the iceberg", *Nature*, 520: 612.

Lyotard, J.-F., 1979. *La Condition postmoderne. Rapport sur le savoir*: Chapter 10. Paris: Minuit.

Macilwain, C., 2015. "The future of science will soon be upon us", *Nature*, 524: 137.

Matthews, D., 2015. "Secret dossier on research fraud suggests government concern over science", *Times Higher Education*, 3 December.

McQuillan, D., 2014. "The Countercultural Potential of Citizen Science", *Media and Communication Journal*, 17(6).

Millgram, E., 2015. *The Great Endarkenment: Philosophy for an Age of Hyperspecialization*. Oxford University Press.

Mirowski, P., 2011. *Science-Mart: Privatizing American Science*. Harvard University Press.

Nature Biotechnology, 2012. "Further Confirmation Needed", Editorial, *Nature Biotechnology*, 30: 806.

Nature, 2015. "Misplaced faith. The public trusts scientists much more than scientists think. But should it?" Editorial, 2 June, *Nature*, 522: 6.
http://www.nature.com/news/misplaced-faith-1.17684

Newman, G., 2015. "9 things to be grateful for about Citizen Science", Monthly Letter from Greg Newman (the CSA board chair) to all CSA members, 12 November.
http://citizenscienceassociation.org/2015/11/12/9-things-to-be-grateful-for-about-citizen-science/

New York Times, 2015. "Scientists Who Cheat". Editorial, 1 June.

Ravetz, J. R., 1971. *Scientific Knowledge and its Social Problems*. Oxford University Press.

Ravetz, J. R., 2011. "Postnormal Science and the maturing of the structural contradictions of modern European science", *Futures* 43: 142–148.

Ravetz, J. R. and Funtowicz, S. O., 2015. "Science, New Forms of", in Wright, J. D., (ed.), *International Encyclopedia of the Social and Behavioral Sciences*, 2nd edition, Vol. 21: 248–254. Oxford: Elsevier.

Ravetz, R. and Saltelli, A., 2015. "Policy: The future of public trust in science", *Nature*, 524: 161.

Saltelli, A. and Funtowicz, S., 2014. "When all models are wrong: More stringent quality criteria are needed for models used at the science-policy interface", *Issues in Science and Technology*, 30(2): 79-85.

Sanderson, K., 2013. "Bloggers put chemical reactions through the replication mill", *Nature*, 21 January, doi:10.1038/nature.2013.12262.

San Francisco Declaration on Research Assessment (DORA). http://www.ascb.org/dora/.

Sarewitz, D., 2015. "Science can't solve it", *Nature*, 522: 413-414.

Springer, 2015. "Springer and Université Joseph Fourier release SciDetect to discover fake scientific papers". https://www.springer.com/gp/about-springer/media/press-releases/corporate/springer-and-universit%C3%A9-joseph-fourier-release-scidetect-to-discover-fake-scientific-papers--/54166

Stiglitz, J., 2010. *Freefall, Free Markets and the Sinking of the Global Economy*. London: Penguin.

Stilgoe, J., 2009. *Citizen Scientists: reconnecting science with civil society*. London: Demos.

Toulmin, S., 1990. *Cosmopolis. The Hidden Agenda of Modernity*. Chicago: University of Chicago Press.

Toulmin, S., 2003 (2001). *Return to Reason*. Harvard University Press.

Van der Sluijs, J. P., 2012. "Uncertainty and Dissent in Climate Risk Assessment: A Post-Normal Perspective", *Nature and Culture*, 7(2): 174-195.

Van Noorden, R., 2015. "Sluggish data sharing hampers reproducibility effort", *Nature News*, 3 June.

Weinberg, A. M., 1961. "Impact of Large-Scale Science on the United States", *Science*, 134 (3473): 161-164.

Whitfield, J., 2012. "Elsevier boycott gathers pace: Rebel academics ponder how to break free of commercial publishers", *Nature*, doi:10.1038/nature.2012.10010.

Wilsdon, J., 2015. "We need a measured approach to metrics", *Nature*, 523: 129, 9 July.

Wynne, B., 1993. "Public uptake of science: a case for institutional reflexivity", *Public Understanding of Science*, 2: 321-337.

Yong, E., 2012. "Nobel laureate challenges psychologists to clean up their act", *Nature News*, 3 October.

2

THE FALLACY OF EVIDENCE-BASED POLICY

Andrea Saltelli and Mario Giampietro

1. Science for policy: predicaments and doubts

The incoming Commission must find better ways of separating evidence-gathering processes from the "political imperative".

In a critique of the usage of science for policy making, Anne Glover, in her capacity as Chief Science Adviser to the President of the European Commission, lamented that evidence-based policy too often turned into its opposite: policy-based evidence. Her counsel, quoted above, was for the European Commission to maintain a more rigorous separation of science from policy (Wilsdon, 2014).

Was this diagnosis correct, and the proposed remedy appropriate? We claim in this chapter that they were not, and furthermore that the problems afflicting the relationship between science and policy run much deeper, with concurrent crises in science, public trust in institutions and sustainability giving rise to a need for new and serious measures.

We will use insights from two main fields of scholarship: science and technology studies (STS) and bioeconomics. From the former field we will make frequent reference to the style of politically engaged science described by Funtowicz and Ravetz as "post-normal science" (PNS) which is pertinent when "facts are uncertain, values in dispute, stakes high and decisions urgent" (Funtowicz and Ravetz, 1991, 1992, 1993). We will discuss questions of epistemic governance, defined as "how knowledge for policymaking should be understood and governed" (Pearce and Raman, 2014).

From the second field of inquiry we shall draw on the seminal work of Georgescu-Roegen and Rosen (Rosen, 1991; Giampietro, 2003), in which the epistemological implications of complexity are addressed at the outset of quantitative analysis.

1.1. *Times ripe with controversy*

In recent years the use of science to inform public policy on various issues has been marked by controversy. Subjects which have provoked great contention and dispute over scientific evidence, as well as intense media coverage, include the impact of pesticides on bees, the necessity of culling badgers, the greenhouse potential of the refrigerant liquid used by Mercedes Benz, the effects of endocrine disruptors, the benefits of shale gas fracking, the fate of children raised by gay parents, the long-term cost of citizenship for immigrants and the benefits of international comparative testing of the educational attainment of children. The highest levels of antagonism have normally been reserved for issues such as anthropogenic climate change and genetically modified organisms (GMOs) – the quintessential 'wicked problems' (Rittel and Webber, 1973). These are issues so deeply entangled in webs of barely separable facts, interests and values that the parties concerned cannot find agreement on the nature of the problem, not to speak of the solution. Kahan (2015) has observed that climate change affects us so intimately that it can define who we are culturally and normatively. We postulate that this kind of intimate relationship, and the culturally-determined attitude toward scientific facts that it engenders, characterizes our interaction with many other phenomena besides climate change and may be at play in all cases where science is called upon to help adjudicate a controversy.

1.2. The crisis in science

The house of science seems at present to be in a state of crisis. A discussion of this crisis can be found elsewhere in this volume (Chapter 1). We only note here that the issues are acknowledged, both in the academic press and the mainstream media, to have become extremely urgent. At the time of writing not a day passes without a comment being registered on some aspect of the crisis—be it reproducibility, peer review, publication metrics, scientific leadership, scientific integrity or the use of science for policy (the last-mentioned provoking the most heated discussion). As discussed elsewhere in this volume (Chapter 1, Chapter 3) the transmission chain of the crisis, from science to scientific advice, is largely a product of the collapse of the dual legitimacy system which was the basis of modernity—that is, the arrangement by which science provided legitimate facts, and policy, legitimate norms.

1.3. Trust, modelling and uncertainty

In a speech before the European Parliament in 2014, Pope Francis sent a blunt 'the Emperor has no clothes' message:

The great ideas which once inspired Europe seem to have lost their attraction, only to be replaced by the bureaucratic technicalities of its institutions. As the European Union has expanded, there has been growing mistrust on the part of

> *citizens towards institutions considered to be aloof, engaged in laying down rules perceived as insensitive to individual peoples, if not downright harmful. (Francis I, 2014)*

The link between this warning and 'science for policy' may not seem obvious, but indeed, rules are the result of policies, and policies are often defended on the basis of 'evidence'. Consider, for example, how science was recruited to support the argument for austerity in public budgets. A ratio of public debt to gross domestic product of 90% was defined by Harvard professors Kenneth Rogoff and Carmen Reinhart as the absolute limit above which growth would be impaired. Debt ratios above this level were thus deemed to be unsafe. A subsequent reanalysis by researchers from the University of Massachusetts at Amherst disproved this finding by tracing it to a coding error in the authors' original work. However, by the time this particular result was repudiated, policies had already been put in place and "[i]n Britain and Europe, great damage has been done as a result" (Cassidy, 2013).

This is but one of the many instances in which improper use of mathematical modelling has been instrumental in justifying flawed policies. Modelling hubris and its consequences are discussed in Saltelli *et al.* (2013) and Saltelli and Funtowicz (2014). In his 2013 work *Never Let a Serious Crisis Go to Waste: How Neoliberalism Survived the Financial Meltdown*, Philip Mirowski devotes a long section (275-286) to the story of how dynamic stochastic general models (DSGE) were the subject of a hearing in the U.S. Senate—"an event in 2010 that was literally unprecedented in the history of economic thought in America" (2013: 275), with sworn testimony by economists such as Sidney Winter, Scott Page, Robert Solow, David Colander and V.V. Chari—to understand how "theorists' tools" had come to be used as policy instruments and why these instruments had been all but useless in anticipating the economic crisis. Queen Elizabeth had a comparable moment of disbelief when questioning British economists on the same subject at the London School of Economics (Pierce, 2008).

Saltelli and Funtowicz (2014) identify several problems in the way mathematical modelling is used to control uncertainty in the process of producing evidence for policy. These include: the rhetorical or ritual use of possibly disproportionate mathematical models to impress or obfuscate; the reliance on tacit, possibly unverified assumptions; the expedient inflation or deflation of uncertainties; the instrumental compression and linearization of the analysis so as to reduce complexity and convey a sensation of prediction and control; and finally, the failure to perform sensitivity analysis, or a merely perfunctory performance thereof.

Sarewitz (2000) offers an original insight into science's loss of authority. For this scholar the problem is not a lack of science, but a surfeit:

> *Rather than resolving political debate, science often becomes ammunition in partisan squabbling, mobilized selectively by contending sides to bolster their positions. Because science is highly valued as a source of reliable information, disputants look to science to help legitimate their interests. In such cases, the scientific experts on each side of the controversy effectively cancel each other out, and the more powerful political or economic interests prevail, just as they would have without the science. (2000: 83)*

2. Vulnerability, uncertainty and governance

In this section we complete our diagnosis of the parallel predicaments of epistemic governance and of science for policy by tracing the roots of the relationship between humans and nature back to early modernity. It is instructive to revisit older or even ancient philosophical and scientific texts, as these are the bedrock in which most present-day thinking is anchored. Not many practicing scientists will have read these sources, but they speak through and to us nonetheless.[1]

2.1. *Science to compensate for human vulnerability: the Cartesian dream*

Humans are reflexive anticipatory systems capable of monitoring and, to a certain extent, predicting events associated with their interaction with the environment (Rosen, 1985: vii). This provided them with a major comparative advantage over other species. The dominion of *Homo sapiens* seems so extensive that the present era has been termed the Anthropocene. As isolated individuals, human beings are vulnerable, and as a species, humans depend on processes beyond their control for their survival. These sources of vulnerability have forced humans to build their identity around organized communities and to adopt communal beliefs. The legitimization of power in a social group is based on the willingness of individuals to trade part of their autonomy, via social contracts, for a reduction in their vulnerability to a host of threats (in the sense of Maslow): that is, for benefits such as personal security against hostile actions by other human beings; food, energy and water security; housing and job security; health care; environmental security; cultural identity and participation. No human society has so far been able to provide protection against death; however, a reduction in the feeling of vulnerability to this event was made possible by religion. In particular, introducing the concept of an eternal life after death attenuated the fear associated with the loss of the physical body, thus affording considerable power to those religions which were able to sustain the claim.

---

[1] We apologize to historians and philosophers of science for the sketchy nature of this summary.

For this reason, societies were for a long time ruled by a combination of religious and military power. With the French revolution and the onset of modernity, the new ruling class replaced religion with science, fulfilling the prophecies of Francis Bacon (1561-1626) and the subsequent formulations of René Descartes (1596-1650) and Nicolas de Caritat, Marquis de Condorcet (1743-1794), to the effect that all social problems would ultimately be solved by knowledge. Bacon's utopia, as described in the "Magnalia Naturae", an appendix to the *New Atlantis*, includes "wonders of nature, in particular with respect to human use":

> *The prolongation of life; The restitution of youth in some degree; The retardation of age; The curing of diseases counted incurable; The mitigation of pain; More easy and less loathsome purgings; The increasing of strength and activity; The increasing of ability to suffer torture or pain; The altering of complexions, and fatness and leanness; The altering of statures; The altering of features; The increasing and exalting of the intellectual parts; Versions of bodies into other bodies; Making of new species; Transplanting of one species into another; Instruments of destruction, as of war and poison; Exhilaration of the spirits, and putting them in good disposition; Force of the imagination, either upon another body, or upon the body itself; Acceleration of time in maturations; Acceleration of time in clarifications; Acceleration of putrefaction; Acceleration of decoction; Acceleration of germination; Making rich composts for the earth; Impressions of the air, and raising of tempests; Great alteration; as in induration, emollition, &c; Turning crude and watery substances into oily and unctuous substances; Drawing of new foods out of substances not now in use; Making new threads for apparel; and new stuffs, such as paper, glass, &c; Natural divinations; Deceptions of the senses; Greater pleasures of the senses; Artificial minerals and cements. (Bacon, 1627: 415-416)*

We leave it to the reader to judge the accuracy of these predictions. One century later Condorcet was so convinced of the potential of physics to solve human predicaments that in the Ninth Epoch of his *Sketch for a Historical Picture of the Progress of the Human Spirit* he states, "All the errors in politics and in morals are founded upon philosophical mistakes, which, themselves, are connected with physical errors" (Condorcet, 1785: 235).

In modern times the main tenets of this line of thinking were resumed by Vannevar Bush at the end of War World II and cogently expressed in the metaphor of the "endless frontier":

> *One of our hopes is that after the war there will be full employment. [...] To create more jobs we must make new and better and cheaper products [...] new products and processes are not born full-grown. They are founded on new principles and new conceptions which in turn result from basic scientific research. Basic scientific research is scientific capital [...] It has been basic United States policy that Government should foster the opening of new frontiers. It opened the seas to clipper ships and furnished land for pioneers. Although these frontiers have more or less disappeared, the frontier of science remains. (Bush, 1945, Chapter 3)*

In the summary of Rommetveit *et al.* (2013):

> *Bacon formulated the basic belief that knowledge gives us power to act in the world to the benefit of our lives. Condorcet elaborated the utopia of a science-based society as one of welfare, equality, justice and happiness. Bush argued that scientific progress and a strong public funding of basic science are necessary conditions to sustain economic growth by the development of new products (or innovation in contemporary vocabulary).*

In spite of the copious amounts of ink spilt by social scientists and STS scholars in warning of the limits of the

Cartesian dream[2], and notwithstanding that 'science wars' were fought between natural and human sciences in the 1980s and 1990s[3], the Cartesian dream is still the prevailing narrative legitimizing scientific practice. Ravetz calls this the "folk-science" of the educated classes:

Indeed, we may say that the basic folk-science of the educated sections of the advanced societies is 'Science' itself in various senses derived from the seventeenth-century revolution in philosophy. This is quite explicit in figures of the Enlightenment such as Condorcet [...] a basic faith in the methods and results of the successful natural sciences, as the means to the solution of the deepest practical problems [...]. (1971: 387)

As discussed by Toulmin in *Cosmopolis: The Hidden Agenda of Modernity*, the vision of 'Cosmopolis' — that is, of a society as rationally ordered as Newtonian physics — took imaginative hold of Western thinking in the 17th century thanks to its extraordinary success in many fields of endeavour and institutionalized an agenda of control, according to which eco- and social systems could be fitted into precise and manageable rational categories. This vision and agenda have survived to the present day, but now more than ever they crash against the complexities of the modern world, endangering the legitimacy of the existing social contracts.

---

[2] Some important texts in this regard include Toulmin's *Return to Reason* (2001) and *Cosmopolis* (1990), Feyerabend's *Against Method* (1975), Lyotard's *The Post-Modern Condition* (1979) and Latour's *We have never been modern* (1993). See the discussion in Chapter 1.
[3] A very concise summary of the 'science wars' can be found in Sarewitz (2000). Wikipedia's entries for 'science wars' and 'two cultures' are also informative. The Canadian Broadcasting Corporation CBC has an excellent series, 'How To Think About Science', with an interview with historian of science Simon Schaffer (see http://www.cbc.ca/ideas/episodes/2009/01/02/how-to-think-about-science-part-1---24-listen/).

2.2. *Reductionism, hypocognition and socially constructed ignorance*

We began this chapter by discussing the evidence-based policy model. In this section we aim to show that this model is based on a series of radical simplifications and linearizations which are congruent with both the Cartesian dream and what has been termed "hypocognition" (Lakoff, 2010) and "socially constructed ignorance" (Ravetz, 1986; Rayner, 2012).

In dealing with complex issues, the application of the Cartesian method of reduction with the goal of isolating linear and direct causal explanations becomes inadequate. Under the Cartesian paradigm, scientific advisers determining fish quotas, for example, will be brought to court if they have set the quota either too high (leading to the collapse of the stock) or too low (entailing loss of income for the fishing community). Seismologists can be sentenced to jail, as happened following the earthquake in Aquila (Italy) in 2009—although the sentence was reversed on appeal—for delivering a reassuring message about the unlikelihood of an earthquake, which was judged *ex post* to have been disproportionate to their limited knowledge of possible future events.

Quantitative analysis is predicated on the selection of a structure—a 'frame'—for approaching a problem. This choice of frame already entails a major compression of the information space that can later be used for governance purposes. The compression operates both at the normative level (through the adoption of a world-view—the choice of the *why*) and at the level of the representation (through the choice of the salient attributes for the description of the system—the choice of the *how*) (Giampietro *et al.*, 2013).

This process is explained by Rayner (2012) in terms of "socially constructed ignorance", which is not the result of a conspiracy but rather of the sense-making processes employed by individuals and institutions:

> *To make sense of the complexity of the world so that they can act, individuals and institutions need to develop simplified, self-consistent versions of that world. The process of doing so means that much of what is known about the world needs to be excluded from those versions, and in particular that knowledge which is in tension or outright contradiction with those versions must be expunged. [...] But how do we deal with [...] dysfunctional cases of uncomfortable knowledge [...]?*

This compression comes at a cost, and, when generalized and institutionalized throughout the system, can lead to the degeneration of a given arrangement, eventually producing a situation of *'ancien régime'*, defined as a state of affairs in which the ruling elites become unable to cope with stressors and adopt instead a strategy of denial, refusing to process either internal or external signals, including those of danger (Funtowicz and Ravetz, 1994).

The compression of the information space results in the exclusion and neglect of knowledge that may be available in established scientific disciplines. Rayner calls these the "unknown knowns", that is, knowledge which exists in academia or society but which is omitted in a given approach. For Rayner (2012), "unknown knowns [are] those which societies or institutions actively exclude because they threaten to undermine key organizational arrangements or the ability of institutions to pursue their goals".

Also ignored due to the compression are the "known unknowns" — gaps and areas of ignorance of which we are aware but which are not considered relevant in the chosen definition of the issue.

The result of this compression is to focus the analysis on a finite set of attributes and goals. This inevitably creates a need for processes of optimization: that is, the ana-

41

lyst is obliged to invest time and energy to find the best solution in the wrong problem space.

Needless to say, the hubris that lies behind this approach increases the fragility of the system, principally in relation to "unknown unknowns", since the optimization process entails a reduction in the diversity of behaviours allowed (because of the elimination of the more poorly performing alternatives) and therefore a reduction in adaptability (because of the neglect of attributes and goals not considered in the optimization). The issue is discussed at length in Taleb's work *Antifragile* (2012).

A lesson from bioeconomics (Giampietro *et al.*, 2012) is that science should address and integrate relevant events and processes that can be observed and described only by accommodating non-equivalent narratives (dimensions of analysis) and different scales (descriptive domains) at the same time. In this framework, the virtues of reductionism (mainly the power to make rational choices based on the clear identification of a finite set of relevant attributes, goals and direct explanations) become a vice. A form of rationality based on a simple problem structure that is applied to a complex issue becomes a "mad rationality" — a concept attributed to social philosopher Lewis Mumford. A good example of this effect is bioethanol derived from corn, in which hundreds of billions of tax-payers' money have been invested only to develop an alternative energy source that consumes more or less the same amount of energy as it produces. In this case a poor framing of the problem, neglecting the full range of implications in order to focus on achieving the stated objectives of improving energy security and reducing emissions, produced a solution which achieved neither of these goals but generated massive profit to powerful lobbies (Giampietro and Mayumi, 2009).

According to Fargione *et al.* (2008), the production of biofuels increases rather than decreases greenhouse gas emissions because of the effect of land-use change—something not considered in the original analysis of the issue. It should be noted that the finding that greenhouse gas emissions from biofuels were comparatively lower than those produced by fossil fuels was a result of simulations using certain models in which biofuels induce food price increases, which in turn entail cuts in food production (Searchinger *et al.*, 2015). In other words, the model that was used to inform policy was based on the assumption that cuts in greenhouse emissions could be achieved at the cost of increased stress on the poor; however, this assumption was not made explicit when this finding was used to identify biofuels as a desirable option.

Unfortunately, the lessons from bioeconomics, based as they are on a richer web of constraints (Giampetro *et al.* 2012), tend to be less reassuring than the messages from the techno-optimists or the eco-modernists[4], which may explain how little attention bioeconomists have received to date.

Socially constructed ignorance can also be defined as the institutional hegemonization of a given mode of story-telling—*i.e.* the pre-analytical choice of a given set of relevant narratives, plausible explanations and pertinent perceptions and representations—which are assumed, by default, to be apt in normative, descriptive and ethical terms to deal with any problem.

This choice may produce situations in which an elephant in the room goes unnoticed, especially if the chosen mode of story-telling has been dressed with a convenient suite of indicators and mathematical models.

---

[4] See http://www.ecomodernism.org/. For a critique, see Monbiot, 2015.

A famous instance of an invisible elephant was the presidential address to the American Economic Association by Nobel laureate in Economics Robert Lucas in 2003 announcing that the "central problem of depression-prevention has been solved" once and for all; another was the 2004 "great moderation" speech of Bernanke, Chair of the U.S. Federal Reserve, about the successful taming of the volatility of business cycle fluctuations. In both cases top-ranking exponents of the ruling paradigm were seemingly unaware of the possibility of financial collapse that would lead to a global economic crisis in following years.

These blunders have fuelled the long-standing dispute over the prevailing paradigm in economics (Reinert, 2008; Mirowski, 2013), and even within the ranks of the discipline new curricula are being studied (INET, 2013). More radically, some voices have called for a reconsideration of economics as the authoritative discipline in the adjudication of social and environmental issues (Ravetz, 1994; Giampietro *et al.*, 2012: 104; Fourcade *et al.*, 2014). In view of the inadequacy of the discipline of economics as a tool to solve socio-economic problems, one could be forgiven for thinking that the discipline had reverted to (or never developed beyond) a state of immaturity. In a chapter entitled "Immature and ineffective fields of inquiry", Ravetz remarks:

> [...] *The situation becomes worse when an immature or ineffective field is enlisted in the work of resolution of some practical problem. In such an uncontrolled and perhaps uncontrollable context, where facts are few and political passions many, the relevant immature field functions to a great extent as a 'folk-science'. This is a body of accepted knowledge whose function is not to provide the basis for further advance, but to offer comfort and reassurance to some body of believers. (1971: 366)*

The irrelevance of inferences produced by heavy modelling tools can often be proved using plain language. To give an example, the aforementioned dynamic stochastic general equilibrium models, DSGE, which are used as policy instruments, do not work if the underlying hypotheses of 'efficient markets' and 'representative agents' are rejected (Mirowski, 2013: 275-286). This is why we insist that framing is closely related to story-telling. There is nothing new in this idea. The benefit of translating the results of mathematical equations into plain language was a teaching of Alfred Marshall in the late nineteenth century (Pigou, 1925: 427), and is not unknown to contemporary economists (Krugman, 2009: 9); still, mathematical models are often used, like Latin, to obfuscate rather than to illuminate (Saltelli *et al.*, 2013).

Note that we are discussing here the use of mathematical models to help produce input for policy—that is, as a tool to generate inferences for policy-making purposes, rather than as scientific method *per se*. For Stiglitz:

Models by their nature are like blinders. In leaving out certain things, they focus our attention on other things. They provide a frame through which we see the world. (2011)

There is nothing wrong in using blinders in the quest for theoretical progress. The problems arise when the same technique is used in the process of designing policy for practical application. That is what Taleb condemns as an attempt to "Platonify" reality (2007: xxix), in a conscious act of hypocognition. Rayner sees it as a strategy to 'socially construct' ignorance, in a movement which he calls "displacement":

[…] displacement occurs when an organization engages with an issue, but substitutes management of a representation of a problem (such as a computer model) for management of the represented object or activity. (2012)

Displacement does not imply imperfect models—which could possibly be improved—but *irrelevant* models, which cannot be corrected through "learning by doing" and hence can do damage for a longer period of time. In this vein, *Financial Times* columnist Samuel Brittain (2011) notes:

> *Nothing has done more to discredit serious economic analysis than its identification with the guesses about output, employment, prices and so on which politicians feel obliged to make. [...] True scientific predictions are conditional. They assert that certain changes [...] will, granted other conditions are met, [...], lead to a certain state of affairs [...]. But they cannot tell us that the required conditions will be fulfilled.*

Evidence-based policy has thus reached a state of apparent paradox, in the sense that certain practices—Rayner's "displacement"—are widely considered to be wrong, but are pursued nonetheless. In this way society is led to associate the stabilization of its own wellbeing with the stabilization of the institutional settings determining the status quo. In the end evidence-based policy becomes an instrument of 'persuasion', defined as what the leadership uses to control attitudes and opinions of the ruled (Chomsky, 2012: 79-80).

2.3. Legitimacy versus simplification

So far the conventional scientific approach to dealing with sustainability issues has been to try to isolate the best course of action by means of deterministic models. This strategy assumes that it is possible to predict the behaviour of complex self-organizing systems (including reflexive systems, such as human societies) and that the quality of the scientific input to the policy process is ensured by the rigour of the methods applied. This assumption overlooks the abundance of uncertainties which—when properly appraised—imply the total inability of these tools to generate useful inferences.

It is futile to expect, for example, that modelling approaches which have failed to predict a purely financial and economic crisis will be able to inform us accurately about the behaviour of a system involving institutions, societies, economies and ecologies. Yet this is what we do when we apply the technique of cost–benefit analysis (CBA) to dimensions of climate change, or claim to estimate the impact on the economy of increased crime rates resulting from hotter temperatures (Rhodium Group, 2014).

Climate processes encompass a multitude of connections in the nexus of energy, water, food and human institutions. The complexity of socio-ecosystems is not a problem to be solved, but an inherent property of self-organizing systems such as ecosystems and human societies. These systems have agency (reproducing themselves and interacting with each other) on various scales. Analysing them one dimension at a time (water, energy, food or land use) and one scale at a time (local, meso or macro) does not work and provides unlimited scope for undesired blows from the law of unintended consequences (Giampietro *et al.*, 2013). When science is used to suppress uncertainty — rather than to explore the sources of our ignorance — failures are likely. These may result in policy failures such as the default option of "more of the same", even if a given policy did not work in the past. Scientific approaches that are forced to operate systematically outside their field of applicability are likely to produce 'bad science' which cannot confidently be used to guide action. This kind of quantitative approach to complex systems can only foster abuse and corruption. As noted by Porter:

> *The appeal of numbers is especially compelling to bureaucratic officials who lack the mandate of a popular election, or divine right. Arbitrariness and bias are the most usual grounds upon which such officials are criticized. A decision made by the numbers (or by explicit rules of some other sort) has at least the appearance of being fair and impersonal. Scientific objectivity thus provides an answer to a moral de-*

mand for impartiality and fairness. Quantification is a way of making decisions without seeming to decide. Objectivity lends authority to officials who have very little of their own. (1995: 8)

3. The solution: establishing new relations between science and governance

The paradigm of evidence-based policy is based on an assumption of the possibility of prediction and control that tends to suppress "scruples" — that is, feelings of doubt or hesitation with regard to the morality or propriety of a given course of action. Our recommendation is to deliberately reintroduce doubts and scruples into the process of deliberation, in a spirit somewhat closer to Montaigne and somewhat further from Descartes, as we will explain below (Toulmin, 1990).

The quality of the process of production and use of scientific information for governance depends on minimizing the negative effect of hypocognition on the final choice of a policy. It is therefore essential to reflect on the process of formalization of the definition of the issue: that is, how the frame was constructed, in semantic terms, and how this selection has cascaded down into a predefined set of data, indicators and mathematical models. In this section we advance some suggestions to aid practitioners.

3.1. Responsible use of quantitative information

A first requirement for the better use of science for policy is the responsible use of quantitative information. This will require the adoption of specific tools of quality control. Practical tools developed in the context of PNS to address these topics are 'NUSAP' and sensitivity auditing:

- NUSAP is a notational system devised for the management and communication of uncertainty in science for policy, based on five categories for characterizing any quantitative statement: Numeral, Unit, Spread, Assessment and Pedigree (Funtowicz and Ravetz, 1990; Van der Sluijs *et al.*, 2005; see also http://www.nusap.net/).

- Sensitivity auditing (Saltelli *et al.*, 2013; Saltelli and Funtowicz, 2014) extends sensitivity analysis as used in the context of mathematical modelling to settings in which the models are used to produce inferences for policy support purposes. Sensitivity auditing questions the broader implications of the modelling exercise, its frame and assumptions, the assessment of the uncertainties, the transparency of the inferences, the veracity of the sensitivity analysis and the legitimacy of the assessment.

3.2. Distinguishing risk from uncertainty

Frank Knight published a very successful book in 1921 entitled *Risk, Uncertainty and Profit* in which he made a distinction between risk, which could be computed mathematically, and uncertainty, which could not. He also observed that:

We live in a world of contradiction and paradox, a fact of which perhaps the most fundamental illustration is this: that the existence of a problem of knowledge depends on the future being different from the past, while the possibility of the solution of the problem depends on the future being like the past. (Knight, 2009: 313)

We suggest a re-learning of Knight's warning, and a more thorough reconsideration of the implications of the difference between risk and uncertainty (Mayumi and Giampietro, 2006). If we ignore this lesson, we will reduce ourselves to figure of the drunkard who searches for his lost keys under a lamppost, even though he knows he lost them elsewhere, because under the post there is light. Following Wynne (1992), we extend the taxonomy of uncertainties to include dimensions of ignorance and indeterminacy:

- Risk — we know the odds.

- Uncertainty — we don't know the odds: we may know the main parameters. We may reduce uncertainty but increase ignorance.

- Ignorance — we don't know what we don't know. Ignorance increases with increased commitments based on given knowledge.

- Indeterminacy — causal chains or networks are open.

For Wynne:

> *Science can define a risk, or uncertainties, only by artificially 'freezing' a surrounding context which may or may not be this way in real-life situations. The resultant knowledge is therefore conditional knowledge, depending on whether these pre-analytical assumptions might turn out to be valid. But this question is indeterminate — for example, will the high quality of maintenance, inspection, operation, etc., of a risky technology be sustained in future, multiplied over replications, possibly many all over the world? (1992)*

As a prerequisite for a more effective use of science for governance, we will need to unlearn some extreme styles of risk analysis in which we compute the odds just because we have a model that allows us to do so.

3.3. Evidence-based policy versus robust policy

We suggest moving away from 'evidence-based policy', based on aggressive but naïve quantification, towards 'robust policy', based on a strategy of filtering of potential policies through rigorous attempts at falsification rather than confirmation. We borrow from Helga Nowotny's (2003) concept of socially robust knowledge: that is, knowledge that has been filtered through the lenses of different stakeholders and normative stances.

The quality check on proposed explanatory narratives and policies should be carried out using the method of falsification with respect to:

- feasibility (compatibility with external constraints);
- viability (compatibility with internal constraints);
- desirability (compatibility with normative values in a given society).

If the policy fails on one of these criteria, we will have identified a bottleneck, a political issue or a true incommensurability. This approach abandons fantasies of prediction, control, planning and optimization, in favour of strategic learning and flexible management.

The approach has elements of similarity with the strategy suggested by Rayner (2012) to overcome socially constructed ignorance: the idea of "clumsy solutions". While socially constructed ignorance helps to keep "uncomfortable knowledge" at bay, clumsy solutions allow it to be processed:

> *Clumsy solutions may emerge from complex processes of both explicit and implicit negotiation. In other words, solutions are clumsy when those implementing them converge on or accept a common course of action for different reasons or on the basis of unshared epistemological or ethical principles [...] They are inherently satisficing [...] rather than optimizing approaches, since each of the competing solutions is optimal from the standpoint of the proposer. Clumsy solutions are inherently pluralistic [...]. (Rayner, 2012)*

The idea of clumsy solutions resonates with the principle of "working deliberatively within imperfections" of the extended participation model of post-normal science and with the "rediscovery of ignorance" advocated by Ravetz (2015: xviii).

A key step in the examination of the feasibility, viability and desirability of understandings of a given problem entails looking at the situation through a variety of lenses — that is, looking at different dimensions and scales of analysis of the problem. The challenges for quantitative analysis are considerable in the area of sustainability (Giampietro *et al.*, 2012, 2014): when examining the feasibility of food security policies against external constraints (a context in which the agricultural system is a kind of black box), we have to measure requirements and supply in terms of kilograms of potatoes, vegetables, animal products, *etc.* However, if we want to study the viability of food security policies in relation to internal constraints (a context in which the black box is the human diet), we have to measure requirements and supply in terms of kilocalories of carbohydrates, proteins and fats. Similarly, when examining the feasibility of energy security policies against external constraints (related to primary energy sources), we have to measure the relevant physical quantities in terms of tons of coal, kinetic energy of falling water, cubic meters of natural gas, *etc.*, whereas if we want to examine their viability in relation to internal constraints (relating to energy carriers), we have to measure the relevant quantities in terms of kilowatt hours of electricity, megajoules of fuels, *etc.* This epistemological predicament is due to the fact that different types of quantitative assessments are non-equivalent and cannot be compressed into a single indicator (Giampietro *et al.*, 2006). Quantitative representations useful to study *feasibility* are not equivalent to those useful to study *viability*, and the information produced by these two typologies of representations cannot be used to study *desirability*, since a discussion of the latter is impossible without involving social actors representing legitimate but often conflicting normative values.

Only after having operationalized these three dimensions for a given sustainability problem is it possible to carry out an informed assessment of possible policy options, with a view to balancing efficiency and adaptability. This can then feed into a multi-criteria characterization of proposed solutions with respect to different normative ingredients.

The proposed approach is equivalent to exploring a multi-dimensional space with a parsimonious and appropriate experimental design, rather than focussing an unrealistic degree of attention on a single point in this space.

3.4. Quantitative story-telling for governance

One normally associates the expression 'story-telling' with a discursive manner of interaction based on imagination, symbolic association, parables and metaphors from literary traditions and the like. By 'quantitative story-telling' we mean the process whereby quantification is run in a context of openness to a variety of 'stories' about what the problem is. Thus the stories themselves become maps, offering orientation for a quantitative analysis of a problem, especially in relation to the existence of constraints, in order to inform the process of decision making. The expression 'story-telling' implies the existence of a 'story-teller': a person, entity or whole society that has to decide on the usefulness of the chosen stories. In our interpretation, the concept of story-telling is more specific than the concept of narrative because it involves a stronger element of agency. The information given in the story will be used by the story-teller to guide action, which in turn provides criteria to verify the quality of the story.

A self-explanatory example of story-telling in relation to food is provided in Figure 1, where the various stories about the 'food chain supply' map onto different perspectives, concerns and dimensions of analysis.

Figure 1. Mapping the food chain supply onto different perspectives

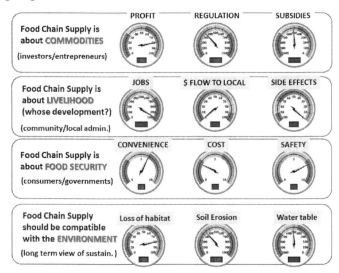

Thus the set of available 'frames' may be widened via quantitative story-telling for governance, with the goal of generating plausible and relevant stories capable of reducing hypocognition in the chosen definition and structuring of the problem—a strategy also suggested by 'cognitive activist' George Lakoff (2014). Quantitative story-telling is also inspired by Ravetz's (1987) concept of "usable ignorance" as an antidote to the "pitfalls in our supposedly secure knowledge or supposedly effective technique".

Quantitative story-telling has the goal of guaranteeing the quality of the chosen story in the given socio-economic and ecological context. The quality of the story-telling moderates the negative effects of the hypocognition associated with the chosen structuring of the problem—the unavoidable neglect of relevant narratives. The fitness of different policy options can then be gauged from the integration of a robust mix of relevant narratives, plausible explanations and pertinent perceptions.

This qualitative check of the coherence of the quantitative information generated by non-equivalent models is essential. In fact, models are by-products of the pre-analytical choice of relevant causal relations, and data are by-products of the pre-analytical choice of relevant perceptions. Confronted with numbers originating in several non-equivalent descriptive domains (logically incoherent quantitative representations), one can no longer rely on the promises of big data and sophisticated algorithms (Lazer *et al.*, 2014). Without a quality check on the chosen mode of story-telling, the accretion of data and models based on unjustified explanations and perceptions will only increase the level of indeterminacy and uncertainty of the results.

The usefulness of the chosen stories needs to be validated using quantitative analysis that must remain coherent across scales and dimensions — that is, a multi-scale, integrated analysis of the functioning of socio-ecological systems, inclusive of their level of openness, for example, to trade. Ignoring these validations steps may imply that relevant aspects of the problems are — perhaps expediently — externalized.

3.5. *Getting the narratives right before crunching numbers*

In his plea for reasonableness versus rationality Toulmin (1990, 2001) contrasts the ideal of Renaissance humanism with the Renaissance scientific revolution. He considers the latter more as a counter-Renaissance, in which the certainties of Descartes supplanted the doubts of Montaigne. In order to return to reason, Toulmin warns, we need to "do the right sums" more than we need to "do the sums right" (2001: 66). This implies a careful selection of the stories to be told before indicators are built, data collected and models run. We need to explore more frames as opposed to selecting just one and filling it with numbers.

We can illustrate this with the persisting controversy surrounding the use of genetically modified organisms, an exemplary wicked problem. The weekly news magazine the *Economist*, discussing a GMO labelling scheme in Vermont (U.S.), commented recently:

> *Montpelier is America's only McDonald's-free state capital. A fitting place, then, for a law designed to satisfy the unfounded fears of foodies [...] genetically modified crops, declared safe by the scientific establishment, but reviled as Frankenfoods by the Subarus-and-sandals set. (2014)*

'Frankenfood' is GMO-based food as defined by its opponents, while the 'Subarus-and-sandals set' is the *Economist's* disparaging allusion to those in Vermont who support a labelling scheme for GMO-based food. The image accompanying the piece shows a hippy-looking public protesting against GMOs. This is a vivid illustration of the cliché: opposition to GMO-based food is generally portrayed as a Luddite, irrational, anti-science position. This is because GMOs are treated as a nutritional 'risk to health' issue, when in fact science has declared GMOs safe for human consumption — the logical conclusion being that society should permit (or even support) their production and consumption.

This frame clashes against the reality of the wider set of citizens' concerns, as measured by, for example, Marris *et al.* (2001). In a list of critical concerns identified through participatory processes, the issue of food safety is conspicuously absent, while others come to the fore:

- Why do we need GMOs? What are the benefits?

- Who will benefit from their use?

- Who decided that they should be developed and how?

- Why were we not better informed about their use in our food, before their arrival on the market?

- Why are we not given an effective choice about whether or not to buy and consume these products?

- Do regulatory authorities have sufficient powers and resources to effectively counter-balance large companies who wish to develop these products?

The variety of frames invoked by these concerns reveals that the prevailing frame, 'safe GMO food versus recalcitrant citizens', is irrelevant for the policy decisions to be taken.

4. Conclusions

The evidence-based policy paradigm should be revised because it fosters radical simplification and compression of representations of issues, explanations and solutions, in a process known as hypocognition. Evidence-based policy cannot be separated from policy-based evidence, with its high reliance upon quantification. The accumulation of data, indicators and mathematical models in support of a given framing of an issue obscures and detracts from the more important task, namely to understand and take into account the implications of the choice of a given frame, bearing in mind that other actors may also act as story-tellers and present different perceptions of the issue to be tackled. A quantitative structuring of a problem allows those that have selected the given mode of story-telling to ignore, through imposed hypocognition, uncomfortable knowledge (Rayner, 2012). Spurious and disproportionate mathematical precision makes it difficult for other practitioners and stakeholders to question the premises of an analysis in plain language. This is not a new finding. In 1986 Langdon Winner was already warning ecologists not to fall into the trap of risk and cost–benefit analyses:

> [T]he risk debate is one that certain kinds of social interests can expect to lose by the very act of entering. [...] Fortunate-

ly, many issues talked about as risks can be legitimately described in other ways. Confronted with any cases of past, present, or obvious future harm, it is possible to discuss that harm directly without pretending that you are playing craps. A toxic waste disposal site placed in your neighborhood need not be defined as a risk; it might appropriately be defined as a problem of toxic waste. Air polluted by automobiles and industrial smokestacks need not be defined as a 'risk'; it might still be called by the old-fashioned name, 'pollution'. New Englanders who find acid rain falling on them are under no obligation to begin analyzing the 'risks of acid rain'; they might retain some Yankee stubbornness and confound the experts by talking about 'that destructive acid rain' and what's to be done about it. A treasured natural environment endangered by industrial activity need not be regarded as something at 'risk'; one might regard it more positively as an entity that ought to be preserved in its own right. (1986: 138-154)

Complex adaptive systems, be they a country's labour market or its forests, are reflexive and continuously becoming 'something else' in order to reproduce themselves (Prigogine, 1980). For this reason it is impossible to predict their future states, because if they manage to reproduce themselves in the long run, they can do so only by moving in a finite time between states that have to be simultaneously:

- feasible—compatible with boundary conditions determined by processes beyond human control;

- viable—compatible with the structure of internal parts and their system of control;

- desirable—compatible with normative values used to legitimize the social contract holding together the social fabric.

In this framework, any definition of what should be considered 'feasible', 'viable' and 'desirable' has to be continuously updated. Changes in processes outside human control will induce changes in boundary conditions (changing the definition of the feasibility domain), in the same way that the full range of consequences of changes taking place in processes which *are* under human control, when spread across scales and dimensions, may impact on both the viability and the desirability of a trajectory, again hampering long-term deterministic predictions.

In human societies, shared normative values are the result of negotiation and modulation of power relations (Lakoff, 2014). This implies that even those modes of story-telling, strategies, and narratives that proved useful for guiding human action in a given historic period may become useless (and therefore potentially misleading) when the meanings attached to the terms 'feasibility', 'viability' and 'desirability' in relation to given goals have changed. To give an example, the "endless frontier" metaphor is less convincing today than it was immediately after World War II. Not only is science now perceived to be an instrument of profit and power, as discussed in Section 1, but the crisis it is suffering has dramatically curtailed its rate of progress (Le Fanu, 2009, 2010). Given this, the acritical perpetuation of dominant narratives may be fatal. In the 19th century Giacomo Leopardi already considered 'Fashion' more deadly than 'Death' (1827).

To give another example from the heart of the contemporary economic debate: the neoclassical economic narrative of perpetual growth based on continuous innovation, which is supposed to reduce inequity through a trickledown effect, was meaningful for developed economies experiencing a period of maximum expansion in their pace of economic activity. As suggested by Daly (1992), the evolution from an "empty world" (a low planetary population) to a "full world" (a large planetary population) means more stringent external constraints and greater environmental impact from human activity. This translates into a reduction in the pace of economic expansion and an increase in inequality within societies — that is, an increase in trouble caused by internal constraints. As an example of what is meant by trouble, recall the debate on the virtues and faults of capitalism *qua* capitalism (Piketty, 2014; Bellamy Foster and Yates, 2014), and the link between inequality and scope for rent-seeking and corruption among elites (Acemoglu and Robinson, 2012).

As any re-adjustment of normative values (requiring large power shifts) is difficult, what role can science play in the necessary refinement of the meanings to be assigned to the concepts of 'feasibility', 'viability' and 'desirability', in different geographical and social contexts?

So far, the role of science for governance and sustainability has been that of a driver of techno-scientification of human progress (see Chapter 3, this volume). As discussed in Section 2, this strategy is an endorsement of the main line of thinking connecting Bacon, Condorcet and Bush, namely that all human problems can be solved by scientific and technological progress. Many institutional actors seem to be mired in the old business of prediction and control, making plans informed by poorly justified scientific results — for example, the results of dynamic stochastic general equilibrium models whose inadequacy we discussed in Section 2, or cost–benefit analyses of the long-term impact of climate on economy and society (Saltelli and d'Hombres, 2010; Rhodium Group, 2014). A disturbing sign of this practice is the impossible precision of the estimates that are churned out: for example, "D.C. climate will shift in 2047" (Bernstein, 2013); or "August 22 was Earth Overshoot Day. In 8 Months, Humanity Exhausted Earth's Budget for the Year" (Global Footprint Network, 2014)[5]. As discussed elsewhere in this volume

---

[5] For a criticism of the latter estimate see Giampietro and Saltelli (2014).

(Chapter 5) the proliferation of digits is a worrying symptom of decaying scientific skills, another element of science's crisis.

Mathematical modelling should no longer have the ritualistic function of divination. The use of the term 'ritual' is not exaggerated. Nobel laureate Kenneth Arrow had an illuminating experience during World War II as a weather officer in the U.S. Army Air Corps working on the production of monthly weather forecasts:

> *The statisticians among us subjected these forecasts to verification and they differed in no way from chance. The forecasters themselves were convinced and requested that the forecasts be discontinued. The reply read approximately like this: "The commanding general is well aware that the forecasts are no good. However, he needs them for planning purposes." (Szenberg, 1992: 47)*

Richard Feynman, in his 1974 commencement address at Caltech, entitled "Some remarks on science pseudoscience and learning how not to fool yourself", calls this type of use of science "cargo cult science":

> *In the South Seas there is a cargo cult of people. During the war they saw airplanes land with lots of good materials, and they want the same thing to happen now. So they've arranged to imitate things like runways, to put fires along the sides of the runways, to make a wooden hut for a man to sit in, with two wooden pieces on his head like headphones and bars of bamboo sticking out like antennas – he's the controller – and they wait for the airplanes to land. They're doing everything right. The form is perfect. It looks exactly the way it looked before. But it doesn't work. No airplanes land. So I call these things cargo cult science, because they follow all the apparent precepts and forms of scientific investigation, but they're missing something essential, because the planes don't land.*

Science has provided incredible comforts to societies in the 'Global North', and many believe that the same benefits are now due to the rest of the world. What is (perhaps wilfully) overlooked by those holding this hope is the fact that the solution of many problems in Northern countries involved externalizing to someone or something else (the environment, future generations or other countries) the negative consequences of the increased level of consumption of resources per capita. Things look quite different when considering the sustainability of technical progress of the whole, interconnected world. At the global level there is no room for externalization — it is a zero-sum game, and there is no 'free-lunch' — someone is paying or will pay for what is consumed. We are dealing with an inextricable confusion of physical, biological, social and ethical issues. Hoping that this problem will be solved by more computer power, more complicated models, bigger databases and more rigour in the scientific method can only result in "cargo cult science". The problem with this type of science is that, as suggested by Feynman, in spite of frenetic activity and all the good intentions in the world, "planes don't land".

References

Acemoglu, D. and Robinson, J., 2012. *Why Nations Fail. The Origins of Power, Prosperity, and Poverty*. Crown Business.

Bacon, F., 1627. "Magnalia Naturae, Praecipue Quoad Usus Humanos".

https://archive.org/details/worksfrancisbaco05bacoiala

Bellamy Foster, J. and Yates, M. D., 2014. "Piketty and the Crisis of Neoclassical Economics", *Monthly Review*, 66(6).

Bernstein, L., 2013. "D.C. climate will shift in 2047, researchers say; tropics will feel unprecedented change first", *Washington Post*, 9 October.

Brittain, S., 2011, "The follies and fallacies of our forecasters", *Financial Times*, May 27.

Bush, V., 1945. "Science: the endless frontier", United States Office of Scientific Research and Development, U.S. Govt. Print Office.

https://www.esci.umn.edu/scitech/courses/HSci3332/vb/VBush1945.html

Carrozza, C., 2015. "Democratizing Expertise and Environmental Governance: Different Approaches to the Politics of Science and their Relevance for Policy Analysis", *Journal of Environmental Policy & Planning,* 17(1): 108-126.

http://dx.doi.org/10.1080/1523908X.2014.914894

Cassidy, J., 2013. "The Reinhart and Rogoff Controversy: A Summing Up", *The New Yorker*, 26 April.

Chomsky, N., 2013 (2012). *Power Systems*. Penguin.

Condorcet, 1785, Marie-Jean-Antoine-Nicolas Caritat, Marquis de Condorcet, *Outlines of an historical view of the progress of the human mind*. See English source at

http://oll.libertyfund.org/titles/1669

Daly, H., 1992. "From empty-world economics to full-world economics: recognizing an historical turning point in economic development", in Goodland, R., Daly, H. E., El Serafy, S. (eds.), *Population, technology, and lifestyle: the transition to sustainability*: 22-37. Washington: Island Press.

Economist, 2014. "Vermont vs science. The little state that could kneecap the biotech industry", 10 May.

Fargione, J., Hill, J., Tilman, D., Polasky, S., Hawthorne, P., 2008. "Land Clearing and the Biofuel Carbon Debt", *Science*, 319(5867): 1235-1238.

Feyerabend, P., 2010 (1975). *Against Method*. London: Verso.

Feynman, R. P., 1974. "Cargo Cult Science. Some remarks on science pseudoscience and learning how not to fool yourself", Caltec's 1974 Commencement Address.

Fourcade, M., Ollion, E. and Algan, Y., 2014. "The Superiority of Economists", discussion paper 14/3, Max Planck Sciences Po Center on Coping with Instability in Market Societies.

Francis I, 2014. "Address of Pope Francis to the European Parliament", Strasbourg, France, 25 November 2014. https://w2.vatican.va/content/francesco/en/speeches/2014/november/documents/papa-francesco_20141125_strasburgo-parlamento-europeo.html

Funtowicz, S., 2006. "What is Knowledge Assessment?", in Guimarães Pereira, Â., Guedes Vaz, S. and Tognetti, S. (eds.), *Interfaces between Science and Society*. Sheffield: Greenleaf Publishers.

Funtowicz, S. and Ravetz, J., 1990. *Uncertainty and Quality in Science for Policy*, Dordrecht: Kluwer Academic Publishers.

Funtowicz, S. O. and Ravetz, J. R., 1994, "Emergent complex systems", *Futures*, 26(6): 568-582.

Funtowicz, S.O. and Ravetz, J. R., 1991. "A New Scientific Methodology for Global Environmental Issues", in Costanza, R. (ed.) *Ecological Economics: The Science and Management of Sustainability*: 137–152. New York: Columbia University Press.

Funtowicz, S. O. and Ravetz, J. R., 1992. "Three types of risk assessment and the emergence of postnormal science", in Krimsky, S. and Golding, D. (eds.), *Social theories of risk*: 251–273. Westport, Connecticut: Greenwood.

Funtowicz, S. O. and Ravetz, J. R., 1993. "Science for the postnormal age", *Futures*, 25 (7): 739–755.

Funtowicz, S. and Ravetz, J. R., 2015. "Peer Review and Quality Control", in *International Encyclopedia of the Social & Behavioral Sciences*, 2nd edition.

Giampietro, M., 2003. *Multi-Scale Integrated Analysis of Agroecosystems*. Boca Raton, Florida: CRC Press.

Giampietro, M., 2008. "The future of agriculture: GMOs and the agonizing paradigm of industrial agriculture", in Guimarães Pereira, Â. and Funtowicz, S. (eds.), *Science for Policy: Challenges and Opportunities*. New Delhi: Oxford University Press.

Giampietro, M. and Mayumi, K., 2009. "The Biofuel Delusion: the Fallacy behind Large-scale Agro-biofuel Production". London: Earthscan Research Edition.

Giampietro, M. and Saltelli, A., 2014. "Footprints to nowhere", *Ecological Indicators*, 46: 610–621.

Giampietro, M., Allen, T. F. H. and Mayumi, K., 2006. "The epistemological predicament associated with purposive quantitative analysis", *Ecological Complexity*, 3(4): 307-327.

Giampietro, M., Mayumi, K. and Sorman, A.H., 2012. *The Metabolic Pattern of Societies: Where Economists Fall Short*. Abingdon: Routledge.

Giampietro, M., Mayumi, K., Şorman, A.H., 2013. *Energy Analysis for a Sustainable Future: The Multi-Scale Integrated Analysis of Societal and Ecosystem Metabolism*. Abingdon: Routledge.

Giampietro, M., Aspinall, R. J., Ramos-Martin, J. and Bukkens, S. G. F. (eds.) 2014. *Resource Accounting for Sustainability Assessment: The Nexus between Energy, Food, Water and Land Use*. Abingdon: Routledge.

Global Footprint Network 2014, "August 22 was Earth Overshoot Day. In less than 8 Months, Humanity Exhausted Earth's Budget for the Year".
http://www.footprintnetwork.org/en/index.php/GFN/page/earth_overshoot_day

Guimarães Pereira, Â. and Funtowicz, S. (eds.), 2015. *Science, Philosophy and Sustainability: The end of the Cartesian dream*. Routledge series: Explorations in Sustainability and Governance.

Institute for New Economic Thinking, 2013. "Institute for New Economic Thinking Launches Project to Reform Undergraduate Syllabus".
http://ineteconomics.org/blog/institute/institute-new-economic-thinking-launches-project-reform-undergraduate-syllabus

Jasanoff, S., 1996. "Beyond Epistemology: Relativism and Engagement in the Politics of Science", *Social Studies of Science*. 26(2): 393-418.

Krugman, P., 2009. *The return of Depression Economics*, New York: W. W. Norton & Company.

Kahan, D. M., 2015. "Climate science communication and the measurement problem", *Advances in Political Psychology*, 36: 1-43.

Knight, F. H., 2009 (1921). *Risk, Uncertainty, and Profit*. Cornell University Library. New York: Hart, Schaffner & Marx.

Lakoff, G., 2010. "Why it Matters How We Frame the Environment", *Environmental Communication: A Journal of Nature and Culture*, 4(1): 70-81.

Lakoff, G., 2014 (2004). *The All New Don't Think of an Elephant: Know Your Values and Frame the Debate*. Revised 2nd edition. Chelsea Green Publishing.

Latour, B., 1993. *We Have Never Been Modern*. Cambridge: Harvard University Press. Originally published 1991 as *Nous n'avons jamais été modernes*. Paris: Editions La découverte.

Lazer, D., Kennedy, R., King, G. and Vespignani, A., 2014. "The Parable of Google Flu: Traps in Big Data Analysis", *Science*, 343: 1203-1205.

Le Fanu, J., 2010. "Science's Dead End", *Prospect Magazine*, 21 July 2010.
http://www.prospectmagazine.co.uk/features/sciences-dead-end

Le Fanu, J., 2009. *Why Us?: How Science Rediscovered the Mystery of Ourselves*. Pantheon.

Leopardi, G., 1827. "Dialogo della Moda e della Morte", in *Operette Morali*. Italian text
http://www.leopardi.it/operette_morali03.php; English version: *Operette Morali – Essays and Dialogues*, 1992, translated by Giovanni Cecchetti. Berkeley: University of California Press.

Lyotard, J.-F., 1979. *La Condition postmoderne. Rapport sur le savoir*, Chapter 10. Paris: Minuit.

Marris, C., Wynne, B., Simmons, P. and Weldon, S., 2001. "Final Report of the PABE Research Project Funded by the Commission of European Communities", Contract number: FAIR CT98-3844 (DG12-SSMI) Dec. Lancaster: University of Lancaster.

Mayumi, K., 2001. *The Origins of Ecological Economics: The Bioeconomics of Georgescu-Roegen*. London: Routledge.

Mayumi, K. and Giampietro, M., 2006. "The epistemological challenge of self-modifying systems: Governance and sustainability in the post-normal science era", *Ecological Economics*, 57: 382-399.

Mirowski, P., 2011. *Science-Mart: Privatizing American Science*. Harvard: Harvard University Press.

Mirowski, P., 2013. *Never Let a Serious Crisis Go to Waste: How Neoliberalism Survived the Financial Meltdown*. Brooklyn: Verso Books.

Monbiot, G., 2015. "Meet the ecomodernists: ignorant of history and paradoxically old-fashioned", *The Guardian,* 24 September.

Nowotny, H., 2003. "Democratising expertise and socially robust knowledge", *Science and Public Policy*, 30(3): 151-156.

Pearce, W. and Raman, S., 2014. "The new randomised controlled trials (RCT) movement in public policy: challenges of epistemic governance", *Policy Sciences*, 47: 387–402.

Pigou, A. C. (ed.), 1925. *Memorials of Alfred Marshall*. London: Macmillan.

Pierce, A., 2008. "The Queen asks why no one saw the credit crunch coming", *The Telegraph*, 5 November.

Piketty, T., 2014. *Capital in the Twenty-First Century*, Belknap Press/Harvard University Press.

Porter, T. M., 1995. *Trust in Numbers. The Pursuit of Objectivity in Science and Public Life*. Princeton University Press.

Prigogine, I., 1980. *From Being to Becoming: Time and Complexity in the Physical Sciences*. San Francisco: W. H. Freeman & Co.

Ravetz, J., 1971. *Scientific Knowledge and its Social Problems*. Oxford University Press.

Ravetz, J. R., 1986. "Usable knowledge, usable ignorance: Incomplete science with policy implications", in Clark, W. C. and Munn, R. (eds.), *Sustainable development of the biosphere*: 415-432. New York: IIASA/Cambridge University Press.

Ravetz, J. R., 1987. "Usable Knowledge, Usable Ignorance, Incomplete Science with Policy Implications, *Knowledge: Creation, Diffusion, Utilization*, 9(1): 87-116.

Ravetz, J. R., 1994. "Economics as an elite folk science: the suppression of uncertainties", *Journal of Post-Keynesian Economics*, Winter 1994-95, 17(2).

Ravetz, J. R., 2015. "Descartes and the rediscovery of ignorance", in Guimarães Pereira, A. and Funtowicz, S. (eds.): xv-xviii.

Rayner, S., 2012. "Uncomfortable knowledge: the social construction of ignorance in science and environmental policy discourses", *Economy and Society*, 41(1): 107-125.

Reinert, E. S., 2008. *How Rich Countries Got Rich . . . and Why Poor Countries Stay Poor*. Public Affairs.

Rhodium Group, 2014. *American Climate Prospectus. Economic Risks in the United States*. Prepared as input to the Risky Business Project, 2014.

Rittel, H. and Webber, M, 1973. "Dilemmas in a General Theory of Planning", *Policy Sciences*, 4: 155–169. Elsevier Scientific Publishing Company, Inc., Amsterdam. [Reprinted in Cross, N. (ed.), 1984. *Developments in Design Methodology*: 135–144. Chicester: J. Wiley & Sons.]

Rommetveit, K., Strand, R., Fjelland, R., Funtowicz, S., 2013. "What can history teach us about the prospects of a European Research Area?", European Union report EUR 26120.

Rosen, R., 1985. *Anticipatory Systems: Philosophical, Mathematical, and Methodological Foundations*. Pergamon Press.

Rosen, R., 1991. *Life Itself: a Comprehensive Inquiry into Nature, Origin, and Fabrication of Life*. Columbia University Press.

Saltelli, A. and d'Hombres, B., 2010. "Sensitivity Analysis Didn't Help. A Practitioner's Critique of the Stern Review", *Global Environmental Change*, 20: 298-302.

Saltelli, A. and Funtowicz, S., 2014. "When all models are wrong: More stringent quality criteria are needed for models used at the science-policy interface", *Issues in Science and Technology*, Winter: 79-85.

Saltelli, A. and Funtowicz, S., 2015. "Evidence-based policy at the end of the Cartesian dream: The case of mathematical modelling", in Guimarães Pereira, A. and Funtowicz, S. (eds.): 147-162.

Saltelli, A., Guimarães Pereira, A., van der Sluijs, J. P. and Funtowicz, S., 2013. "What do I make of your Latinorum? Sensitivity auditing of mathematical modelling", *International Journal of Foresight and Innovation Policy*, 9(2-4): 213–234.

Sarewitz, D., 2000. "Science and Environmental Policy: An Excess of Objectivity", in Frodeman, R., 2000. *Earth Matters: The Earth Sciences, Philosophy, and the Claims of Community*: 79-98. Prentice Hall.

Searchinger, T., Edwards, R., Mulligan, D., Heimlich, R., Plevin, R., 2015. "Do biofuel policies seek to cut emissions by cutting food?", *Science*, 347(6229): 1420-1422.

Shapin, S. and Schaffer, S., 2011(1985). *Leviathan and the Air-Pump: Hobbes, Boyle, and the Experimental Life*. Princeton: Princeton University Press.

Stiglitz, J. E., 2011. "Rethinking macroeconomics: what failed, and how to repair it", *Journal of the European Economic Association*, 9(4): 591–645.

Szenberg, M. (ed.), 1992. *Eminent Economists: Their Life Philosophies*. Cambridge: Cambridge University Press.

Taleb, N. N., 2007. *The Black Swan. The Impact of the Highly Improbable*. London: Penguin.

Taleb, N. N., 2012. *Antifragile: Things That Gain from Disorder*. Random House.

Toulmin, S., 1990. *Cosmopolis. The Hidden Agenda of Modernity*. Chicago: University of Chicago Press.

Toulmin, S., 2003 (2001). *Return to Reason*. Harvard University Press.

Van der Sluijs, J. P., Petersen, A. C., Janssen, P. H. M., Risbey, J. S. and Ravetz, J. R., 2008. "Exploring the quality of evidence for complex and contested policy decisions", *Environmental Research Letters*, 3 024008 (9pp).

Van der Sluijs, J., Craye, M., Funtowicz, S., Kloprogge, P., Ravetz, J. and Risbey, J., 2005. "Combining Quantitative and Qualitative Measures of Uncertainty in Model based Environmental Assessment: the NUSAP System", *Risk Analysis*, 25(2): 481-492.

Wilsdon, J., 2014. "Evidence-based Union? A new alliance for science advice in Europe", *The Guardian*, 23 June.

Winner, L., 1989 (1986). *The Whale and the Reactor: a Search for Limits in an Age of High Technology*. Chicago: University of Chicago Press.

Wynne, B., 1992. "Uncertainty and environmental learning: reconceiving science and policy in the preventive paradigm", *Global Environmental Change* 2(2): 111-127.

3

NEVER LATE, NEVER LOST, NEVER UNPREPARED

Alice Benessia and Silvio Funtowicz

Introduction

In this chapter we examine innovation as a dynamic system of forces that constantly and necessarily redefine the boundaries between science, technology and the normative sphere of liberal democracy. We consider innovation as a phenomenon which is on a path-dependent trajectory, with origins in the scientific revolution and the emergence of the modern state in the 16th and 17th centuries. We give an overview of its evolution through the lens of the 'demarcation problem': that is, we consider innovation with respect to the boundaries that demarcate scientific research from other human activities. More specifically, we look at *how* those boundaries have been drawn over time, *by whom*, and *to what ends*. In this exploration, we identify three main modes of demarcation that function as principles and drivers of innovation, defining the structure of the space in which it evolves: we call these 'separation', 'hybridization' and 'substitution'.

'Separation' refers to the ideal division between the facts of science and the values of governance and to the corresponding 'dual system of legitimacy' that regulates the 'modern' relation between knowledge and power. In this framework, uncertainty and complexity are ideally externalized from the realm of scientific knowledge and activity. 'Hybridization' corresponds to the transition from curiosity-motivated 'little' science to big, industrialized science, in which science and technology, discovery and invention, and facts and values are blended (hybridized) in 'technoscientific' endeavours. In this framework, uncertainty and complexity cannot be effectively banished; however, they can be reduced and ideally controlled through quantitative risk assessment and management. 'Substitution' involves the replacement of natural resources with technoscientific artefacts, of decision making with data management and of understanding with making. Ultimately, substitution leads to the replacement of science itself by technology in a process that defines and legitimizes (*i.e.* demarcates) innovation. Values are substituted by facts, in the sense that normative issues are reduced to technical matters that can supposedly be resolved by technoscientific means. In this framework, uncertainty and complexity are acknowledged, managed and ideally eliminated.

These modes of demarcation have emerged consecutively and are presented here in an historical perspective, but, as we will see in our study of the narrative of innovation, they also co-exist to various degrees[1]. As the story unfolds, we will introduce a frame of reference for the narrative of innovation, providing examples of how various technoscientific innovations in

---

[1] Indeed, different historical accounts are possible, emphasising continuity and parallel developments rather than distinct phases in the style of scientific research and application. See, for example, Crombie (1994).

the fields of nanotechnology, space exploration and emer-
emergent information and communication technology
(ICT) have been represented to consumers, investors and
governments in such a way as to affirm their epistemic,
normative and economic legitimacy. Considered
collectively, these examples illustrate how innovation is
constituted by appealing to the ideals of separation,
hybridization and substitution and how the main
proponents of innovation—scientists, administrators and
entrepreneurs—draw the boundaries of their territories in
order to ensure their survival and expansion (or, in
modern terms, competitiveness and growth). These
examples shed some light on the current complex and
contentious relationships between science, technology and
governance.

In the final section of this chapter we will perform a
thought experiment: we will assume that all doubts
regarding the impact of technoscience have been laid to
rest, that the risks associated with it have been mitigated
and that its promises have been fulfilled. This assumption
will allow us to imagine what kind of world is implied,
based on what values and with what implications for
whom. This exercise will make plain some of the main
contradictions inherent in the prevailing narrative of
innovation and will point to possibilities for developing
alternative narratives.

Separation: science as representation of the true and good

Narratives of progress can be construed as
demarcating strategies—that is, rhetorical repertoires that
legitimize certain worldviews and systems of knowledge
and power. As such, all narratives imply a specific set of
relationships between science, technology and the
normative sphere of liberal democracy.

In the early stages of the scientific revolution and the modern state in the mid-17th century, we find the emergence of a dual legitimacy system, ideally separating the objective world of facts—the realm of science—from the subjective world of human affairs and values, regulated by emerging institutions of governance. The banishment of uncertainty and complexity from the jurisdiction of scientific endeavour was essential for this dual system to function: the object of scientific investigation had to be protected from the inner world of the experimenter (ruled by subjective sensations, emotions and passions) and from the world outside the laboratory (governed by social and political values)[2].

In this framework, science had to be dissociated not only from ideology (meaning metaphysics, religion and politics), but also from *technology*. Justification, discovery and knowing were deliberately distinguished from application, invention, and making, in an effort by scientists to compete with engineers and religious authorities for epistemic legitimacy and material resources.

In a compelling account of science in Victorian England, sociologist Thomas Gieryn reconstructed the demarcating strategies of John Tyndall, successor to Michael Faraday as Professor and then Superintendent at the Royal Institution in London, in charge of delivering lectures demonstrating the relevance and the progress of scientific knowledge to both lay and scientific audiences (Gieryn, 1983, 1989). In Tyndall's view, science could compete with religion on the grounds of being practically useful, empirically sound, sceptical with respect to any authority other than the facts of nature, and free from

---

[2] Galileo Galilei performed this separation in the realm of science by distinguishing between primary and secondary qualities of objects in *The Assayer* (1623).

subjective emotions. Confronted with the practical suc-successes of Victorian engineering and mechanics, he de-described science as a fount of knowledge on which technological progress depended. It thus had to be repre-represented as theoretical and systematic in the search for causal principles and laws, and as perfectly disinterested. Furthermore, science was to be understood as a means to culture. The genuine ambivalence of Tyndall's boundary work between scientific and social institutions was a product of the inherent tensions between basic and applied research and between the empirical and the theoretical aspects of inquiry in the 19th century.

In Tyndall's wake, the demarcation of science as an analytical problem preoccupied and even dominated the endeavours of philosophers of science, driven by different ideological commitments but all searching for essential properties that could demarcate science as a unique and privileged source of knowledge (Ravetz, 1991). In the tradition of the Vienna Circle of the 1930s, in their struggle against the dogma and metaphysics of clerical forces, science was the unique path to human truth and improvement, and the inductive method, based on repeated observations and experiments, was considered to be the only foundation for making general statements about nature.

Cognizant of the limits of empirical induction as a method for scientific investigation, Karl Popper invoked the moral quality of "daring to be shown wrong" and made it the core of a new approach based on the principle of "falsifiability". If a theory could not in principle be refuted (*i.e.* 'falsified') by empirical data, it was not scientific (Popper, 1935). In Popper's view, refutation could immunize science against all sorts of pseudo-scientific activities (such as socialism and psychoanalysis) emerging from the collapse of authority in central Europe after World War I.

In the early 1940s, in opposition to "local contagions of anti-intellectualism which could become endemic" (*i.e.* the rise of various forms of fascist and nationalist movements), the American sociologist Robert Merton expressed a need for a new "self-appraisal" of scientific practice and knowledge, noting that the faith of Western culture in science was, in Veblen's words, no longer "unbounded, unquestioned and unrivalled". In his essay "The normative structure of science", Merton attributed to modern science a unique ability to provide "certified" knowledge, thanks to the institutionalization of distinctive social norms in the scientific community, in the form of a specific ethos that drove progress (Merton, 1973/1942). The ethical and epistemic value of science ensured by the Mertonian norms of communalism, universalism, disinterestedness and organized scepticism helped to delimit a "republic of science" — an autonomous community of peers, self-governed though shared knowledge and under no form of authority other than knowledge itself (Polanyi, 1962; Merton, 1968).

The ideal of separation between facts and values; the suppression of complexity in favour of certainty and objectivity; the identification of moral virtue with epistemic value and meaning; and the uniquely privileged position of scientific knowledge: these are the elements of a foundational narrative of scientific knowledge and investigation that defines the inherited approach to science for policy, in which science should "speak truth to power", providing neutral and objective evidence to support rational decisions in the form of logical deductions (Wildavsky, 1979). As we will see, this mode of demarcation of science is still invoked today in various ways, despite the radically different conditions in which it is applied and the growing conflict over the dual legitimacy system. A paradigmatic illustration of the persistence of this demarcation model was given when Professor Anne Glover, at the time Chief Scientific

Adviser to the President of the European Commission, recommended that the incoming Commission find "better ways of separating evidence-gathering processes from the political imperative" (Wilsdon, 2014), as discussed elsewhere in the present volume (Chapter 2).

Hybridization: technoscience for growth, power and prosperity

Advancing along the narrative of innovation to the American post-World War II context, we find a different set of boundaries and balance of forces in play, provoking a shift in the modern ideal of science and the emergence of new demarcating principles. In his 1945 report "Science, the Endless Frontier", the first American presidential science adviser Vannevar Bush affirmed the primacy of basic scientific research as the engine of economic growth:

> To create more jobs we must make new and better and cheaper products. We want plenty of new, vigorous enterprises. But new products and processes are not born full-grown. They are founded on new principles and new conceptions, which in turn result from basic scientific research. Basic scientific research is scientific capital. Moreover, we cannot any longer depend upon Europe as a major source of this scientific capital. Clearly, more and better scientific research is one essential to the achievement of our goal of full employment. (Bush, 1945)

Bush's thesis is that the work of individual scientists as they pursue truth in their laboratories ultimately contributes to the common good by feeding into the technological development that stimulates economic growth. Bush evokes Tyndall's definition and legitimation of science as a source of knowledge for technological development. However, crucially, in this view science and technology no longer compete with each other for epistemic authority and material resources; rather, they become intimately related and jointly instrumental to the common goals of the production of goods and the creation of jobs. It was the early stage of a new type of modernity, based on the hybridization of science and technology in the name of technoscientific progress and its promise of unlimited wealth and prosperity.

In this process, 'science-based' technology is granted the epistemic and moral legitimacy of science and it becomes the incarnation of the Cartesian dream of power and control over nature. When newly elected American President Dwight D. Eisenhower gave his lecture on "Atoms for Peace" in 1953, the development of nuclear weapons was told as the first technoscientific story of emancipation, in the form of a promise that nuclear power would provide unlimited energy to people and nations (Eisenhower, 1953). The *New York Times* of 17 September 1954 reports this vision in a speech by the Chairman of the U.S. Atomic Energy Commission, Lewis Strauss:

Our children will enjoy in their homes electrical energy too cheap to meter [...] will travel effortlessly over the seas and under them and through the air with a minimum of danger and at great speeds, and will experience a lifespan far longer than ours, as disease yields and man comes to understand what causes him to age.

Technology thus became a source of wonders and unlimited possibilities, and science developed into "the art of the soluble" (Medawar, 1967); it became a 'normal', disenchanted, puzzle-solving profession, as described in the widely acknowledged work of Thomas Kuhn (Kuhn, 1962).

Early signs of a general transition from curiosity-oriented science, with its object of creating universal knowledge, to big, industrialized technoscience, with the function of producing corporate know-how, were given in 1961 in Eisenhower's "Farewell Address to the Nation":

Today, the solitary inventor, tinkering in his shop, has been overshadowed by task forces of scientists in laboratories and testing fields. In the same fashion, the free university, historically the fountainhead of free ideas and scientific discovery, has experienced a revolution in the conduct of research. Partly because of the huge costs involved, a government contract becomes virtually a substitute for intellectual curiosity. For every old blackboard there are now hundreds of new electronic computers.

In the course of this process of hybridization, the relationship between science, technology and society changed. As laboratories became testing grounds and applied science expanded into the real world, the inherently hybrid notion of 'safety' entered the scene, calling into question the values of the psychological, social, political and economic spheres and the facts of science.

The republic of trans-science

In 1962 marine biologist Rachel Carson published a volume about the possible side effects of pesticides. Evoking a distressing scenario in which nature would awake from winter without any bird to celebrate it, Carson's book *Silent Spring* fostered the emergence of the American environmentalist movement, triggering public awareness and concerns about the potentially devastating drawbacks of the chemical heroes of the Green Revolution and the fight against malaria. In his 1967 book *Reflections on Big Science*, American nuclear physicist Alvin Weinberg, administrator of the Oak Ridge National Laboratory during and after the Manhattan Project, cast doubt on the safety of civilian nuclear technology (Weinberg, 1967). A few years later, in an essay for a meeting of the American Nuclear Society, he described the wonders of nuclear energy in terms of a "Faustian bargain" that would demand unprecedented new forms of vigilance and longevity (stability and long-term commitment) in social institutions (Weinberg, 1994). In 1972, while studying the biological effects of exposure to low-level radiation, he took a further step towards recognition of the transformation taking place within science and technological development: in a landmark article in the journal *Minerva*, he proposed a principle of demarcation for a new class of problem that he called "trans-scientific" and which was emerging as a consequence of big science (Weinberg, 1972):

> *Many of the issues which arise in the course of the interaction between science or technology and society – e.g., the deleterious side effects of technology, or the attempts to deal with social problems through the procedures of science – hang on the answers to questions which can be asked of science and yet which cannot be answered by science. I propose the term trans-scientific for these questions since, though they are, epistemologically speaking, questions of fact and can be stated in the language of science, they are unanswerable by science; they transcend science. In so far as public policy involves trans-scientific rather than scientific issues, the role of the scientist in contributing to the promulgation of such policy must be different from his role when the issues can be unambiguously answered by science.*

Trans-science essentially breaks down the ideal separation between the facts of science and the values affecting policy decisions. The "republic of trans-science", in Weinberg's terms, has elements of both a political republic and a republic of science. The rights of its citizens are succinctly captured by Weinberg in the saying, "He whose shoe pinches can tell something to the shoemaker": this was possibly the first time the concept of 'stake-holder' was applied in this context.

It then became important to know how to demarcate scientific questions, which could be dealt with exclusively within the protected walls of Mertonian science, from trans-scientific ones, which required an opening of the gates. Moreover, the distinction itself was, of course, not a matter of experimental science.

At the same time, as "every old blackboard" was being substituted with "hundreds of new electronic computers", the world of statistical systems analysis was discovered, once again pushing scientific research out of the laboratories, this time into the world of computer simulations. Harvey Brooks, solid-state physicist and administrator at Harvard, was one of the pioneers of this transition, as a member of the International Institute for Applied Systems Analysis (IIASA) since its foundation in 1972 and later chair of its U.S. Committee for more than a decade. In a letter to *Minerva*, in the same issue in which Weinberg coined the term trans-science, Brooks pointed out that understanding the evolution of complex systems governed by large classes of non-linear equations, which are at the heart of simulation models, was also a trans-scientific challenge, as it could not be addressed by science alone (Brooks, 1972). In the same year, the Club of Rome published the report "The limits to growth" (Meadows *et al.*, 1972). Based on the so-called World3 system dynamics model for computer simulation, the essay explored for the first time the global trans-scientific issue of how exponential demographic and economic growth interact with finite resource supplies. It was the beginning of the sustainable development movement.

In the transition to big science, not only did the boundaries of the republic of science become fuzzy and permeable; its inner structure, supposedly based on objectivity and neutrality, also proved to be questionable. The autobiographical account of the race for the discovery of DNA, published in 1968 by James Watson, exposed the highly intellectual and affective personal dimension to scientific research, revealing that bitter competition and acrimonious dispute were more nearly the rule in science than the exception (Watson, 1968). The influence of nei-neither the inner, subjective world of emotions and passions nor the outer world of social, political and economic values could be ignored in the practice of science, as illustrated by Bruno Latour in *Science in Action* (1987), using this very example.

In 1974 American sociologist Ian Mitroff published the results of an extensive study performed at NASA, the heart of another U.S. 'big science' project: the space exploration programme in the race for the Moon. Based on a substantial set of interviews with a selected group of Apollo moon scientists, Mitroff uncovered the existence of a deep-seated ambivalence among the researchers with respect to the putative norms of science. The Mertonian norms supposedly underpinning the curiosity-motivated ideal of science were dynamically balanced by corresponding counter-norms such as particularism (versus universalism), solitariness (versus communism), interestedness (versus disinterestedness), and organized dogmatism (versus organized scepticism). The balancing of norms and counter-norms was instrumental to surviving in large technoscientific enterprises characterized by hierarchical systems and high economic and political stakes, and this skill defined a new model of entrepreneurial technoscientist. A few years later, the Bayh-Dole Act of 1980 institutionalized this model by authorizing private ownership of inventions financed by federal funding.

Towards the risk society

Uncertainty and complexity cannot be effectively externalized from the realm of technoscientific endeavour. They emerge in the interaction of technoscience with the real world of social and ecological systems and in the interplay between the individual and organizational dynamics of big enterprises. The modern ideal of science 'speaking truth to power' had to be adjusted to control for this new configuration of forces. If uncertainty and complexity could not be suppressed, they had to be operationalized, statistically controlled (by science), and openly discussed (by parliamentary democratic processes), in order for the dual system of legitimacy to be preserved. The notion of 'risk', which could be technically assessed and managed by scientific experts and exploited to speak (a probabilistic) truth to power, was an unsuccessful attempt to solve this emerging tension.

The 1979 nuclear disaster of Three Mile Island was the first prominent example of a 'trans-scientific' failure, in which technological breakdown was inextricably entangled with organizational and management malfunction. The event prompted sociologist Charles Perrow to define as "normal accidents" the inevitable, built-in vulnerability to collapse of tightly coupled, highly complex technological systems, such as nuclear plants (Perrow, 1984/1999).

In 1985, during his second term as Administrator of the U.S. Environmental Protection Agency (EPA), William Ruckelshaus admitted that many of the EPA's regulations depended on the answers to questions that could be asked of but not answered by science — that is, the EPA was dealing in the regulation of trans-scientific problems (Ruckelshaus, 1985). It was the beginning of the so-called "risk society", as defined in 1986 by sociologist Ulrich Beck in a work that treated the growing awareness that the goods and bads of technoscientific development were two sides of the same coin and that risks were woven into the very fabric of technoscientific progress (Beck, 1986/1992).

It was not only sociologists and public officials, but also natural scientists, who had to learn to deal with the risks and ambiguities of technoscientific enterprise and the new boundaries being traced along the trajectory of progress. In 1986, on a cold winter morning a few months before the nuclear catastrophe at Chernobyl, the NASA Space Shuttle *Challenger* exploded a few seconds after take-off, live on national television. In the aftermath of the accident, theoretical physicist and Nobel laureate Richard Feynman was called on to examine the causes of the disaster as a member of the Presidential Commission in charge of the investigation (later known as the Rogers Commission).

Following his investigation, Feynman famously recounted, again on national television, the physical causes of the event: the lack of resilience and breakdown at low temperatures of an O-ring seal in one of the rocket boosters, due to faulty design, which caused a fatal leak of pressurized burning gas. However, in his minority report for the Commission (Appendix F to the main report), Feynman examined the causes at a different level[3], questioning the evaluation of safety and the risk assessment procedures within NASA. In his "personal observations on the reliability of the Space Shuttle", Feynman pointed out that the probabilities of failure—the risk of a fatal accident for the *Challenger*—were matters of "opinion" at NASA, ranging from roughly 1 in 100 in the estimate of the working engineers, to 1 in 100,000 in the evaluation of the management. A difference of this magnitude can only be explained in two ways. First, the managers of the project may have deliberately underestimated the risks, effectively lowering the safety standards to ensure the timely execution of the scheduled mission (and consequently the continuous supply of funds). This seems plausible, given that President Ronald Reagan was due to give his State of

---

[3] Feynman's move to a higher level of organization in the search for the causes of the accident can be interpreted as a significant attempt to overcome the limits of the reductionist approach, within and outside the boundaries of the physical sciences. For an interesting account of this perspective see Fjelland (2015).

the Union address to the United States Congress on the day of the launch—a national technoscientific success would have been an outstanding achievement. The second possible explanation was an "almost incredible lack of communication" between NASA officials and engineers, due to the complexity and inefficiency of the Agency's governance structure. In either case, the causes of the *Challenger* disaster were to be traced to the inherent ambiguities and inconsistencies (the interplay between norms and counter-norms, in Mitroff's terms) in the political environment and in the organizational structure of the responsible institution.

Interestingly, in the conclusion of his Appendix, Feynman refers to the "reality" of natural laws, which "cannot be fooled" by human interests, thus essentially appealing to the possibility and even the necessity of separating the facts of science from the values of decision-making (and giving facts the priority), in the name of technological safety. As an inquisitive, curiosity-motivated commissioner in charge of a public investigation, Feynman recognized the complexities and ambiguities of hybridized technoscience, but still fell back on the option of retreating behind the lines of Mertonian science to ensure that science remained the representation of both the True and the Good[4]. He effectively acted as a bridge between the first phase of modernity, based on the demarcating principle of separation, and another, involving the blending of science, technology and society (hybridization).

---

[4] In his renowned lecture on "Cargo Cult Science", Feynman argued vigorously for a falsifiable science and for the moral commitment of scientists to do their best to falsify their own work, following a tradition of demarcation from Popper to Merton (Feynman, 1974). See excerpts in this volume (Chapter 2).

The Scanning Tunneling Microscope and the demarcation of nanotechnology: observing and manipulating

In parallel to growing tensions between science, technology and society with respect to safety, a vigorous demarcating effort was being made by the new entrepreneurial scientists to secure the material conditions and epistemic authority of their endeavours and outputs. Technoscientific development was promoted as a source of power and control over natural phenomena. A few months after the *Challenger* disaster, the 1986 Nobel Prize in Physics was awarded to three scientists: Ernst Ruska, Gerd Binnig and Heinrich Rohrer. One half of the Prize went to Ruska "for his fundamental work in electron optics, and for the design of the first electron microscope" — work which was actually done in the early 1930s.

The other half went jointly to Binnig and Rohrer "for their design of the Scanning Tunneling Microscope" (STM), an evolution of the first electron microscope, capable of imaging individual atoms and bonds with a resolution up to 100 times higher than its predecessor. What is noteworthy is that the three physicists were not honoured for discovering new physical laws or phenomena, but for the invention of new fundamental tools for the visualization of the atomic world, developed for and patented by private companies (Siemens and IBM). In their acceptance speech, Binnig and Rohrer effectively define and legitimize (*i.e.* demarcate) their invention by skilfully navigating the ambiguities of hybridized technoscientific development. While describing the technical aspects of their instrument, they repeatedly emphasized the beauty and the wonder of atomic surfaces, appealing to the modern ideal of the scientist as an explorer of unknown territories, epitomized by figures such as Galileo and Robert Hooke. At the same time, they effectively evoked the technological power and heroism of space exploration, by transforming the arid diagrams of scanned atomic structures into black and white staged photographs of actual physical models, suggesting remote planetary surfaces (Figures 1 and 2)[5].

Figure 1. Surface studies by scanning tunneling microscopy

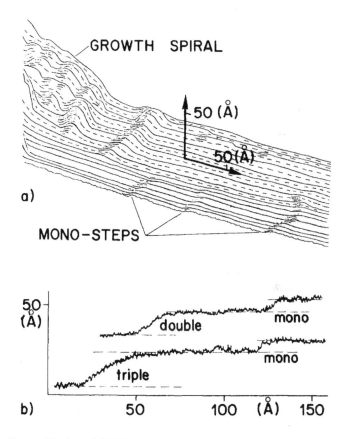

Source: Binnig *et al.* (1982).

[5] See Nordman (2004) for an account of the relation between the narrative of nanotechnology and space exploration.

Figure 2. The relief

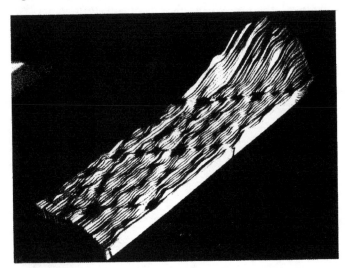

Source: Binnig and Rohrer (1986).

Observing and intervening are inherently coupled at the atomic scale, given the dominance of quantum mechanical laws, so the ability to determine the position of individual atoms with controlled precision would require a further technological advance. Binnig and Rohrer address it towards the very end of their Nobel lecture, and Richard Feynman is invoked once again:

> *Besides imaging, the STM opens, quite generally, new possibilities for experimenting, whether to study nondestructively or to modify locally [...] and ultimately to handle atoms and to modify individual molecules, in short, to use the STM as a Feynman Machine. (Binnig and Rohrer, 1986)*

The "Feynman Machine" is an explicit reference to a talk about the possibilities of miniaturization that Feynman gave in 1959 at the California Institute for Technology, entitled "There is plenty of room at the bottom". In that talk, he essentially advocated a fundamental shift from a reductionist model privileging the use of theoretical, mathematical language to describe and understand the book of nature, to an instrumental and applied reductionism based on the development of new technologies for observing and manipulating matter at the atomic level.

> *We have friends in other fields – in biology, for instance. We physicists often look at them and say, "You know the reason you fellows are making so little progress?" (Actually I don't know any field where they are making more rapid progress than they are in biology today.) "You should use more mathematics, like we do." They could answer us – but they're polite, so I'll answer for them: "What you should do in order for us to make more rapid progress is to make the electron microscope 100 times better." […] The problems of chemistry and biology can be greatly helped if our ability to see what we are doing, and to do things on an atomic level, is ultimately developed – a development which I think cannot be avoided (Feynman, 1959).*

As a celebrated theoretical physicist who had given a Nobel lecture in the very same room in 1965 in respect of the discovery of quantum electrodynamics (QED), Feynman was ideally positioned to confer the epistemic and moral authority of Mertonian science on the new technoscientific endeavour of the STM. As a visionary figure bridging old and new phases of modernity, he functioned as a credible, propelling force for the demarcation efforts of the newly recognized nanotechnologists. In fact, Feynman's symbolic role was so effective that his 1959 talk was retroactively 'discovered' and became the foundational narrative of the field of nanotechnology[6].

---

[6] Carefully planned and coordinated for more than a decade by the engineer Mihail C. Roco, the delineation of the field of nanotechnology culminated with the announcement by President Bill Clinton of the first federal government programme for nanoscale research and development projects, defined as the National

Meanwhile, the "Feynman Machine" became a reality in 1990, in the hands of another pair of IBM scientists, Don M. Eigler and Erhard K. Schweizer. Their achievement was announced simultaneously on the cover of *Nature* (Eigler and Schweizer, 1990) and the *New York Times* (Browne, 1990), with another iconic hybrid image (Figure 3), working at once as a representation of experimental scientific evidence—a number of xenon atoms purposefully arranged on a layer of nickel at extremely low temperature—and as a demonstration of corporate power—IBM conquering matter at its very core[7].

Figure 3. Cover of *Nature* 344(6266), 1990

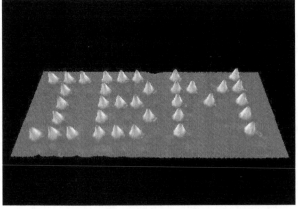

Nanotechnology Initiative, at the California Institute of Technology in 2001 (McCray, 2005).

[7] The image conjures up the American flag on the surface of the Moon, in a display of national power and in celebration of the American victory in the Cold War space race. Even more interestingly, it calls to mind the gesture of Hiram Maxim, the inventor of the first portable automatic machine gun in Victorian England, who shot the letters V.R. ("Victoria Regina") into a wall in the presence of the Queen, to demonstrate the military potential of his invention.

Precaution and post-normal science

While the atomic logo of IBM signalled a triumph of
the Cartesian ideals of power and control, awareness of
the possible unforeseen consequences of technoscientific
development continued to grow. The public and political
acknowledgement that "nature cannot be fooled" and that
the modern ideal of separation of facts and values had to
be adjusted in view of the pathologies of technoscientific
progress predicated the United Nations Conference on
Environment and Development (UNCED) in Rio de
Janeiro in 1992. Also known as the "Earth Summit", the
conference coincided with the emergence of the sustainable development movement. Principle 15 of its official
statement, the "Rio Declaration on Environment and Development", introduced a political mode of demarcation,
based on the notion of precaution:

> *In order to protect the environment, the precautionary ap-
> proach shall be widely applied by States according to their
> capabilities. Where there are threats of serious or irreversible
> damage, lack of full scientific certainty shall not be used as a
> reason for postponing cost effective measures to prevent en-
> vironmental degradation. (United Nations, 1992)*

The precautionary approach introduced the idea that science can be *temporarily* unable to produce a conclusive
and exhaustive body of knowledge fit to serve as a basis
for rational decision making. Certainty about the future
consequences of action was substituted with quantitative,
statistically manageable uncertainty — that is, with risk
assessment and cost–benefit analysis. Through this "taming of chance" (Hacking, 1990), uncertainty was officially
recognized as a third value, along with truth and falsity,
in the realm of possible scientific outcomes.

Uncertainty was accepted, however, only as a temporary state of knowledge which was bound to shift sooner
or later to one or the other value on the true/false scale.

Meanwhile, a political choice would have to be made in order to minimize the risk of harm to people and the environment, even if it implied a potential economic loss. The "endless frontier" of science-based technological progress and growth could and had to be temporarily circumscribed, until new predictive certainty could be achieved. The implicit assumption in this model is that any lack of knowledge can be reduced with time, resources and more computational power, leaving untouched the modern relationship between the truth of objective scientific knowledge and the good of rational, evidence-based decision-making. In other words, the precautionary principle can be interpreted as a technical fix to alleviate the conflict within the dual legitimacy system, without modifying its underlying assumptions.

Around the same time, a new mode of demarcation was proposed in the philosophical work of Silvio Funtowicz and Jerome Ravetz, under the label of "post-normal science" (Funtowicz and Ravetz, 1993). This approach held that uncertainty around a technoscientific issue cannot be treated as an independent variable and linearly reduced, as in the conventional approach associated with the precautionary principle. Rather, it must be understood to be closely related to the stakes involved and to be governed by highly non-linear, trans-scientific dynamics. When the stakes are low, as in confined laboratory science, the correlation is less evident, and uncertainty can be externalized with no visible effects; when the stakes are high, as they are in big technoscientific projects, the correlation is pronounced, and the consequences of disregarding uncertainty can be severe. In this perspective, the facts of science and the values underlying decision-making processes cannot be separated, and the decision-making process must be opened up to the participation of "extended peer communities" (De Marchi, 2015).

In tracing the trajectory of innovation along its course from science and technology we have witnessed a progression in three stages: first, curiosity-motivated science, competing with technology, metaphysics and the realm of human affairs for epistemic authority and material resources; second, basic science as 'scientific capital', underpinning technological development for prosperity and growth; lastly, big technoscientific enterprise, entailing inherent risks and drawbacks and the inevitable intermingling of facts and values. With the last shift, uncertainty and complexity cannot be fully and explicitly externalized, and conflicts consequently arise within the dual legitimacy system of the contemporary state.

At this stage, two main ways forward appear on the horizon. One corresponds to a commitment to abandon the delusive modern ideal of separation: this would be the continuation of the trajectory from trans-science to postnormal science. The other, representing the institutional and corporate reaction, focuses on implementing measures to contain the tensions, in order to preserve, and even reinforce, the modern power divide: this is the trajectory of quantified, operationalized uncertainty and complexity, based on (more or less) precautionary risk assessment and management. As we will see below, the current dominant narrative of innovation follows this latter path.

Substitution: innovation for growth and survival

Starting with the attack on the heart of the American financial system on 11 September 2001, the first decade of the new millennium was characterized by a growing awareness of systemic crisis, with economic, social, political and environmental components. Climate change, biodiversity loss, resource scarcity, the rise of terrorist movements and political instability became public and urgent concerns to be addressed on a global level; in 2008, a financial meltdown hit the U.S. economy and propagated to the European Union, triggering the worst global economic crisis since the Great Depression of the 1930s.

In 2010, against this backdrop, Máire Geoghegan-Quinn was appointed Commissioner for Research, Innovation and Science of the European Commission—a post previously denominated "Commissioner for *Science and Research*". This shift, with science slipping quietly to the end of the title, corresponded to the advent of a new demarcating narrative, in which the term 'innovation' took the place, quite literally, of technoscientific development, not only as a source of growth, prosperity and social good, but also as a salvific solution to the ongoing crisis.

The naming of innovation as the engine of economic growth, social prosperity and environmental sustainability was the last semantic manoeuvre in a powerful and highly articulated narrative of progress intersecting with the trajectory of sustainability (Benessia and Funtowicz, 2015). Within this coevolving path, society has been asking science and technology to fulfil (at least) three essential functions: to increase or at least to sustain our wellbeing; to preserve us from the possible adverse consequences of our acting towards this goal; and to manage those adverse consequences or unfavourable circumstances, should they arise. The unchallenged economic policy aims of growth, productivity and competitiveness, reinforced by the globalization of the economy, are fundamental aspects of this relationship with science. In effect, if we accept these goals as a given for improving and extending human welfare on this planet, then we (continue to) set ourselves the paradoxical ambition to sustain a steady increase in global resource consumption within a closed, finite system with limited stocks and bio-geo-chemical resilience (Rockström *et al.*, 2009; Elser and Bennett, 2011; see the discussion in Chapter 1, this volume).

The situation is becoming even more complex, as both the technological and ideological lock-ins of our life-support systems present us with a double-bind, quite painfully clear in the wake of the 2008 financial collapse: we cannot keep moving indefinitely along our current trajectory — but not doing so would jeopardize the economic prospects not only of future generations, but also, decidedly, of our own.

The narrative of innovation offers a repertory of potential solutions to this paradoxical situation. In particular, it counsels us to take into account an essential hidden variable, which Malthus proverbially overlooked: even though natural supplies may be limited, human creativity is *unlimited* and so is the potential to decouple growth from scarcity, improving efficiency in the use of natural resources and ultimately *substituting* them altogether with substantively equivalent, technologically optimized artefacts. At the same time, innovation is invoked to control and even eradicate complexity, uncertainty and the risk of failures through the implementation of effective *ad hoc* technoscientific fixes. The Cartesian ideals of power and control which were at the root of the transition to techno-scientific hybridity have become instruments of economic and even of human survival.

In the European Union strategy for the second decade of the century, innovation is considered instrumental to achieving and nurturing "smart, sustainable and inclusive growth" (European Commission, 2010a). It is furthermore named as the "only answer" (European Commission, 2010b) to some of the world's most pressing societal challenges: "combatting climate change and moving towards a low-carbon society" (European Commission, 2011a) and managing the problems of "resource scarcity, health and ageing" (European Commission, 2010b). The principles of the so-called 'green economy' and the Ecomodernist Manifesto, published by the Californian Breakthrough

Institute, provide other poignant, exemplary instantiations of this Promethean approach (Breakthrough Institute, 2015; Lewis, 1992).

In addition, innovation is cast as the mainstream solution to the problem of sustaining growth in a hyper-saturated market, with its potential to open up new avenues of competition and consumption and to populate them with new jobs and ever more seductive products and services. One of the seven flagship initiatives designed and launched to deliver on the objectives of the European Union's 2020 Strategy is the "Innovation Union", "aiming to improve framework conditions and access to finance for research and innovation so as to ensure that innovative ideas can be turned into products and services that create growth and jobs" (European Commission, 2010a: 3).

To all intents and purposes, this set of arguments is a reformulation of Vannevar Bush's ideals of science-based technological development for growth and prosperity, but, interestingly, the words 'science' and 'scientific' rarely figure in this discourse. Rather, economic growth and new jobs are produced by "research and innovation" which are transformed into "innovative ideas". In essence, the demarcating strategy is the same, but the object to be demarcated is different and vaguer. Moreover, the context in which the narrative unfolds is radically changed. In the post-World War II period, the American people were ready to welcome the great expansion of production with the enthusiasm of a new-born culture of mass consumption. The horizon of resource scarcity and environmental degradation was still far away. Moreover, in a period of peace, and while Europe lay in ruins, the USA could rely only on itself and the "endless frontier" of scientific and technological development.

By contrast, in the race for market share that characterizes the early 21st century, European technoscientific development has to withstand the pressures of the global market:

> *We need to do much better at turning our research into new and better services and products if we are to remain competitive in the global marketplace and improve the quality of life in Europe. (European Commission, 2010a)*

The immediate post-war challenge to emerge and expand has by now turned into a struggle for economic survival. Sustaining growth requires competitive technoscientific innovation. In a lecture given in Brussels with the eloquent title "Winning the innovation race", Commissioner Geoghegan-Quinn made reference to this pressure (Geoghegan-Quinn, 2012):

> *There is no shelter for un-competitive firms or economies. Competitiveness is the new law of economic gravity, which no one can defy.*

Further, only innovation can bear the weight of this law:

> *And now it is knowledge and ideas that drive competitiveness, not tangible assets.*

The knowledge and the ideas evoked here are clearly still anchored to the worldview of Vannevar Bush. Once again, however, 'science' is completely absent from the stage: the term is not used by the Commissioner in her speech, other than to refer to the life and social sciences. This omission presages the beginnings of a significant new transition from 'science-based' technology and big, industrialized technoscience to a fragmented, broader ideal of creative research at the service of market-oriented technology. This embryonic new form of scientific research is related to the Victorian ideal of the practitioner/gentleman amateur, today embodied by the individual entrepreneur/do-it-yourself (DIY)/citizen/garage scientist (Ravetz and Funtowicz, 2015).

Finally, a crucial assumption must hold for this narrative to be viable: citizens of developing, developed and declining economies have to value and ultimately buy, both metaphorically and literally, the processes and products of technoscientific innovation. This means that societal expectations of the products have to be stimulated, and concerns about their ills deflected (European Commission, 2013; ESF, 2009).

In the words of Geoghegan-Quinn, in a short video interview at the Lisbon Council in 2010:

Innovation means that we bring all the wonderful scientific research that we have, all the way along a chain, until we get it into products, we sell it on the market. We develop products and create products that the markets are there for, and the people will want to buy. That is, at the end of the day, how we can develop research to retail. (Geoghegan-Quinn, 2010)

To sum up, innovation can now be defined as a process of creative (scientific) research that leads to the production of new technologies that sustain growth and ensure survival: through the optimization and the substitution of our natural resources, the creation of new goods and jobs, and the deployment of suitable silver bullets, protecting us from the complexity of socio-ecological problems as they emerge.

Technology, sustainability, growth and science thus comprise a constellation of dynamic forces in a space with mutable and ambiguous boundaries. To better understand the emergence and development of the current dominant narrative of innovation, we will focus now on how these forces have been operating and how the corresponding boundaries have been drawn. As we will see, new demarcating strategies emerge from these complex dynamics, based on a principle of substitution.

Smarter planets and the demarcation of the Internet of Things: decision making and data management

In the autumn of 2008, in the middle of the financial storm, the U.S. multinational company IBM launched one of its most ambitious global campaigns, based on the idea of building a "smarter planet"[8]. On 8 November, a few days after the election of Barack Obama to the U.S. Presidency, IBM Chairman and CEO Sam Palmisano presented his narrative of smart innovation in a fifteen-minute speech at the U.S. Council of Foreign Affairs (Palmisano, 2008). In his talk, the planet as a whole was described as a single, highly complex and interconnected socio-technical system, running at a high and increasing speed and demanding more and more energy and resources; climate, energy, food and water needed to be efficiently managed in order to meet the challenges of a growing population and a globally integrated economy; a number of sudden and unexpected wake-up calls such as the crisis in the financial markets had to be recognized as the signs of a dangerous fracture that had to be controlled; the leaders of both public and private institutions had to acknowledge this radical change and seize the opportunities offered by technoscientific innovation to "change the way in which the world works" (Palmisano, 2008). The planet was thus conceived of as a complex machine that would cease to function if not manipulated with the appropriate technological tools.

No sooner had the crisis scenario been presented than IBM's demarcating narrative of innovation moved straight to the resolution: namely, that we *already* have the technological power and control to turn our predicament into an opportunity. As the boundaries of our finite, physical world become more evident in the transition to an era of

---

[8] IBM's "Let's build a smarter planet" campaign by Ogilvy & Mather won the 2010 Gold Effie Award in marketing communications.

resource scarcity, this narrative imagines a technoscientific transition to an apparently boundless universe of digdigital information, virtual connectivity and computational power, allowing us to optimize our way of living and become efficient enough to sustain increased consumption. The three fundamental axes of the new technological revolution are articulated by the terms 'instrumented', 'interconnected' and 'intelligent', which in combination define the notion of 'smart' and, in the context of the European Union, describe the so-called Internet of Things[9]. *Instrumented* reflects the indefinite proliferation and diffusion of the fundamental building block of the digital age, the transistor (up to one billion per human at the infinitesimal cost of one ten-millionth of a cent). As all these transistors become *interconnected*, anything can communicate with anything else. In this vision, we can monitor and *control* our planet with unprecedented precision and capillarity by causing the realms of the physical, the digital and the virtual to converge. Finally, everything can become *intelligent*, as we are able to apply our ever-increasing computational power to sensors, end-user devices and actuators, in order to transform the ocean of data that we collect into structured knowledge and subsequently into action.

Palmisano portrays this transition not only as possible and desirable, but also as required and urgent, both to prevent further collapse of our life-support systems and to sustain competitiveness in the global market:

> *It's obvious, when you consider the trajectories of development driving the planet today, that we're going to have to run a lot smarter and more efficiently – especially as we seek the next areas of investment to drive economic growth and to*

---

[9] The Internet of Things is defined as a dynamic global infrastructure of networked physical and digital objects augmented with sensing, processing and networking capabilities (Vermesan *et al.*, 2011).

> *move large parts of the global economy out of recession [...].*
> *These mundane processes of business, government and life –*
> *which are ultimately the source of those 'surprising' crises –*
> *are not smart enough to be sustainable. (Palmisano, 2008)*

The implicit assumption in this speech is, of course, that the tools for new, *smarter* leadership required are techno-scientific and that IBM can deliver them.

The technoscientific narrative of a corporate marketing initiative depends intrinsically on the function of selling goods, as products and services, and might therefore not be considered representative of a deeper political, economic, cultural and existential transition. However, on the path-dependent trajectory of innovation, the same demarcating strategies can be found in private companies' plans for market share expansion and in public institutions' long-term engagements for the future, as both sectors are engaged in cultivating and surviving the overarching model of competitiveness and consumption growth[10]. It is the case in the EU 2020 strategy for "smart, sustainable and inclusive growth", which incorporates the Internet of Things pathway in one of its key Flagship Initiatives, the "Digital Agenda". In a three-minute video by the European Commission Directorate General for Information Society and Media, we find one of the characters expressing her concerns about energy management as follows:

> *It's crazy that we doubled our use of energy in the last fifty*
> *years. We can't keep this up. If we want to be smart about*
> *energy, we should let energy be smart about itself. (Euro-*
> *pean Commission, 2012)[11]*

---

[10] In this sense, the difference between public and private becomes marginal as in both cases the subject of the demarcating narrative is not a product to be promoted, but a specific *kind of world* in which the proposed innovation is the only possible sustainable option.

[11] Female character no.1.

In this framework, leaders of firms, cities and nations are responsible *only* for choosing the most effective means of technoscientific optimization, in order for the system at stake to govern itself in the most efficient way. In other words, a radical shift is taking place in the dual system of legitimacy and its balance of forces, as political 'power' moves from reliance on scientific 'truth' as the basis of rational decisions, to delegating control over both the True and the Good to automatized technoscientific tools.

Three framing epistemic and normative assumptions need to be in place in order for this demarcating narrative to function. First, it must be accepted that the inherent complexity of the interaction between socio-ecological and technological systems can be reduced to a measurable set of simplified structured information. Second, the required 'facts' have to be equated with supposedly relevant data, filtered through the appropriate information technologies. Third, the *quality* of the decision-making processes must be completely independent of the normative sphere of values — a move which requires sufficient computational power to distinguish data from noise and to assign them a meaning that can transform them into an operationalized notion of knowledge. This overall scenario represents a transition from the ideals of separation and hybridization to a new demarcating strategy based on a principle of substitution, in which the normative sphere of politics and decision-making on public policy issues is reduced, hybridized and ultimately supplanted (substituted) by a technoscientific regime of data analysis and management.

Even more fundamentally, it is not only the issues which demand decision making that are transformed and reduced, but also the 'we' concerned by those issues. Indeed, the ultimate consequence of this set of assumptions is that the most effective decision-maker is in fact the fusion of a physical, a virtual and a digital being: a cyborg or a robot. IBM's supercomputer Watson, a "deep question answering" (DQA) machine, which outsmarted its predecessor Big Blue by winning the U.S. TV game *Jeopardy!* is a clear, early incarnation of this idea (Thompson, 2010)[12].

Palmisano ended his 2010 speech at the Royal Institute of Foreign Affairs in London with these words:

> *Let me leave you with one final observation, culled from our learning over the past year. It is this: Building a smarter planet is realistic precisely because it is so refreshingly non-ideological. (Palmisano, 2010)*

The epistemic, normative and ultimately metaphysical framework of efficiency for smart and sustainable growth is presented by Palmisano as a modern, inevitable consequence of progress for the common good. If our world is a slow, obsolete and congested socio-technical machine ruled by the laws of thermodynamics rather than by those of governance, then (the promise of) technoscientific innovation to optimize its functioning becomes an objective necessity.

Conclusion

In this journey along the trajectory of innovation, we began by looking at science in the early phase of modernity: an oligarchic, exact, objective and uniquely privileged form of knowledge which should remain *separate from* the world of values and human affairs. We then transitioned into the phase of big, industrialized technoscience, in which science was *hybridized with* human affairs as a strategy to secure growth, prosperity and profit. Finally, we entered into a recent third phase, based on a principle

---

[12] Watson is conceived of and proposed as the best instrument to decide in highly complex and urgent situations, ranging from financial transactions to clinical and diagnostic decisions and the management of mass emergencies.

of substitution, in which science *becomes* a human affair, defined as innovation: the unbounded, automatized tool for enhancing, treating and rescuing our slow, congested, analogue world.

Bearing in mind this progression, if we now fully accept the assumptions and promises of this last phase and imagine that all the issues regarding the inherent risks and pathologies of technoscience have been settled, we can reflect on the implications of this narrative of innovation: what kind of world is signified, populated by whom and with what consequences? Such a reflection will help to illuminate possible alternative trajectories and narratives.

Let us begin by revisiting the narrative of innovation proposed by the former CEO of IBM, essentially anticipating the EU Digital Agenda by two years.

In this perspective, we are compelled to logically deduce that the "mundane processes" of our professional, political and private lives have to be technologically enhanced (to become 'smart') in order to avoid a collective crash of the system. The crises we are facing are not at all surprising: they are caused by our own inability to cope with the overall complexity of the processes manifest in our world. Moreover, as we have seen, this technological upgrade is not only logically required, but also feasible and, above all, desirable, as it optimizes our ways of living, making life easier and happier.

However, if we look more closely at the implications of this demarcating narrative of innovation, a number of inherent contradictions emerge. First, the very same technologies that are designed to help us deal with complexity actually generate more: the intricate patterns of interactions and demands of this world, which we can supposedly manage only with the aid of ICT, are intensified by the real-time pervasiveness of the ICT itself. In practice, we are being provoked to run faster and faster by technologies that were intended to help us catch up with ourselves. In addition, referring back to the ideas of Charles Perrow about the consequences of high complexity and tight coupling, we might deduce that this transition inevitably makes us existentially more fragile and vulnerable to technoscientific failures.

Second, if we fully embrace the technological upgrade and agree to delegate the management of the mundane processes of our lives to connected machines, then we are acquiescing to the idea that we should live in a world of happiness, in which we are never late, never lost and, most of all, never unprepared. This world would be a place in which every minute of our lives would need to be virtually controlled and functionally oriented. In other words, we *cannot* be late, lost or unprepared. It is a world, therefore, in which our relationship with the unknown is tacitly eliminated. This form of technological eradication of uncertainty entails renouncing one of the fundamental sources of human creativity and learning: our capacity to adapt to complexity and the unexpected (Benessia *et al.*, 2012). This in turn implies a new contradiction, intimately related to the first: what seems to make us safer and more efficient may be the cause of heightened vulnerability to change.

In this scenario, regardless of the initial conditions of our personal values, expectations and desires, the dynamics of our 'un-smart' and 'messy' planet compel us to delegate both our knowledge and our agency to the required technoscientific power and to embrace and creatively contribute to the accompanying inner transformation of living beings[13].

---

[13] A fully analogous set of arguments can be articulated in relation to the technological platform of synthetic biology. See Benessia and Funtowicz (2015).

If we take these narratives seriously, the ICT-based so-cial transformation becomes simply inevitable and moves beyond the limits of democratic discussion. More gener-ally, if we revisit the main framework of the demarcating narrative of innovation, we find the same inevitability: there is no reason to collectively discuss the proposed technological transition, as we are supposed to want it, need it and be able to have it (Benessia and Guimarães Pereira, 2015). Inherently normative concerns are reduced to technical issues, and their technical solutions are framed in terms of economic feasibility, risk mitigation and public acceptance. In this sense, the democratic foun-dation of social and political action is replaced with the merely procedural coordinates of an essentially *win-win* scenario. Once again, in Palmisano's words:

Building a smarter planet is realistic precisely because it is so refreshingly non-ideological. (Palmisano, 2010)

This reminds us of another key passage of Eisen-hower's "Farewell Address to the Nation" (1961):

The prospect of domination of the nation's scholars by Fed-eral employment, project allocations, and the power of money is ever present – and is gravely to be regarded. Yet, in holding scientific research and discovery in respect, as we should, we must also be alert to the equal and opposite dan-ger that public policy could itself become the captive of a sci-entific-technological elite.

A possible bearing for new narratives would be to chal-lenge the inevitability of the current technoscientific tra-jectory of innovation and to collectively explore the normative space of values and political options, investi-gating the actual feasibility and desirability of the emer-gent technology platforms, in relation to *what kind* of world we want to sustain and *for whom*.

Indeed, the ultimate fate of any innovation funda-mentally depends on identifying *what* the goods and the bads actually are and *for whom*, at any given time. The quality of a technological innovation is a function of the underlying driving forces and how its effects are valued. If we deconstruct the dominant framing of innovation, we may find a collective democratic space to discuss different criteria of quality. For example, do we think that it is feasible and desirable to give up diversity, individuality and our relationship with the unknown in the name of efficiency and functionality, or to subordinate living to functioning? Do we believe that it is the only possible solution for our current predicament?

More generally, the question becomes: what categories are needed to describe *what* needs to be transformed and *how*? *Who* decides on the definitions to be adopted for the various categories? Reflecting on these questions makes it possible to explore alternative trajectories for innovation and to redefine the criteria to assess its quality[14]. Robust and resilient innovations can only emerge from opening up the collective space of options for both the framing of the problems to be resolved and the tools proposed to resolve them. This process will require reflection on our relationship with life-supporting infrastructures and processes and with the other living beings (including humans) that we implicitly include or exclude when we say 'we'.

In light of these considerations, the historical exploration of the trajectory of innovation that we have undertaken becomes an instrument to foster awareness of where we find ourselves along its path, so that we might collectively choose whether and how to intervene to modify its dynamics. As we have seen, terms like science, technology, democracy, ideology and sustainability are con-

---

[14] For an account of how this approach can be applied to the case of biotechnology for food production, see Benessia and Barbiero (2015).

stantly being redefined and re-legitimized along the path, using various demarcating strategies, for various purposes, in various contexts.

For example, if we consider the democratization of science from the point of view of subscribers to the prevailing narrative of innovation, we might be inclined to value its potential to increase public engagement and participation. However, if we look at the same issue through the lens of our narrative of demarcation, we might realize that what is being democratized is a specific, normatively fixed ideal of scientific research and practice, predicated on the eradication of complexity and applied to an equally specific, normatively fixed and mechanically standardized and optimized ideal of living. Being more aware of this constant process of demarcation and redefinition might allow us to develop new tools to understand where we are actually heading and to open up a democratic space for the plotting of alternative routes.

Acknowledgements

We are grateful to Ragnar Fjelland for adding the perspective of A.C. Crombie.

Part of this research was developed at the request of the Joint Research Centre of the European Commission, under Expert Contract 530023 of 1 September 2014.

References

Beck, U., 1986. *Risikogesellschaft*. Frankfurt: Suhrkamp. (English translation 1992). *Risk society: towards a new modernity*. London: Sage.

Benessia, A. and Barbiero, G., 2015. "The impact of genetically modified salmon: from risk assessment to quality evaluation", *Visions for Sustainability* 3: 35-61.

Benessia, A. and Funtowicz S. O., 2015. "Sustainability and technoscience: what do we want to sustain and for whom?", *The International Journal of Sustainable Development*, Special Issue: In the Name of Sustainability, 18(4): 329-348.

Benessia, A. and Guimarães Pereira, Â., 2015. "The Dreams of the Internet of Things: do we really want and need to be smart?", in Guimarães Pereira, Â. and Funtowicz, S. (eds.), 2015. *Science, Philosophy and Sustainability: The end of the Cartesian dream*: 79-99. Routledge series: Explorations in Sustainability and Governance. New York: Routledge.

Benessia, A., Funtowicz, S. O., Bradshaw, G., Ferri, F., Ráez-Luna, E. F. and Medina, C. P., 2012. "Hybridizing sustainability: Towards a new praxis for the present human predicament", *Sustainability Science* 7(1): 75-89.

Binnig, G., H. Rohrer, Ch. Gerber, and E. Weibel, 1982. "Surface Studies by Scanning Tunneling Microscopy," *Physical Review Letters* 49(1): 57-60.

Binnig, G. and Rohrer, H., 1986. Nobel Lecture. http://www.nobelprize.org/nobel_prizes/physics/laureates/1986/binnig-lecture.pdf

Breakthrough Institute, 2015. "An ecomodernist manifesto." http://www.ecomodernism.org

Brooks, H., 1972. "Letters to the Editor (Science and Transscience)". *Minerva*, 10(2): 323-329.

Browne, M. W., 1990. "2 Researchers Spell 'I.B.M.' Atom by Atom", *New York Times*, 5 April.

Bush, V., 1945. "Science, the endless frontier". United States Office of Scientific Research and Development, U.S. Govt. Print Office.

Carson, R., 1962. *Silent Spring*. Boston: Houghton Mifflin.

Crombie, A. C., 1994. *Styles of scientific thinking in the European tradition: the history of argument and explanation especially in the mathematical and biomedical sciences and arts*. London: Duckworth.

De Marchi, B., 2015. "Risk Governance and the integration of scientific and local knowledge", in Fra. Paleo, U. (ed.), *Risk Governance. The Articulation of Hazard, Politics and Ecology*: Chp. 9: 149-165. Berlin: Springer.

Eigler, D. M. and Schweizer, E. K., 1990. "Positioning Single Atoms with a Scanning Tunneling Microscope", *Nature*, 344: 524-526.

Elser, J., and Bennett, E., 2011. "A broken biogeochemical cycle", *Nature*, 478: 29-31.

European Commission, 2010a. "EUROPE 2020: A Strategy for Smart, Sustainable and Inclusive Growth." Communication from the Commission, COM(2010)2020.

European Commission, 2010b. "Europe 2020 Flagship Initiative: Innovation Union." Communication from the Commission, COM(2010)2020.

European Commission, 2012. "Internet of Things Europe - The movie: Imagine everything was linked..." https://www.youtube.com/watch?v=nDBup8KLEtk.

European Commission, 2013. "Science for an informed, sustainable and inclusive knowledge society." Policy paper by President Barroso's Science and Technology Advisory Council, Brussels, 29 August.

Feynman, R., 1959. "There's plenty of room at the bottom". California Institute of Technology.

Feynman, R., 1986. "Personal Observations of the reliability of the Shuttle". Appendix F to the Report of the Presidential Commission on the Space Shuttle *Challenger* Accident, Vol. 2. NASA. http://history.nasa.gov/rogersrep/v2appf.htm

Feynman, R. P., 1974. "Cargo Cult Science: Some remarks on science, pseudoscience, and learning how not to fool yourself". California Institute of Technology, 1974 commencement address.

Fjelland, R., 2015. "Plenty of room at the top", in Guimarães Pereira, Â. and Funtowicz, S. (eds.), 2015. *Science, Philosophy and Sustainability: The end of the Cartesian dream*: 13-25. Routledge series: Explorations in Sustainability and Governance. New York: Routledge.

Funtowicz, S. and Ravetz, J., 1993. "Science for the post-normal age", *Futures*, 31(7): 735-755.

Funtowicz, S. and Ravetz, J., 1994. "Emergent Complex Systems". *Futures*, 26(6): 568-582.

Geoghegan-Quinn, M., 2010. In "What is innovation?" Participants in the Lisbon Council's 2010 Innovation Summit give
their answers to the meaning of innovation. The Lisbon
Council 2010.
http://www.youtube.com/watch?v=2NK0WR2GtFs

Geoghegan-Quinn, M., 2012. "Winning the innovation race." The
2012 Robert Schuman Lecture, the Lisbon Council.
http://www.youtube.com/watch?v=O_PiChA0swo

Gieryn, T. F., 1983. "Boundary-work and the demarcation of
science from non-science: Strains and interests in professional ideologies of scientists", *American Sociological Review*,
48(6): 781-795.

Gieryn, T. F., 1999. *Cultural boundaries of science*. Chicago: University of Chicago Press.

Hacking, I., 1990. *The taming of chance*. Cambridge, UK: Cambridge University Press.

Kuhn, T. S., 1962. *The structure of scientific revolutions*. Chicago:
University of Chicago Press.

Latour, B., 1987. *Science in Action*. Harvard University Press

Lewis, M. W., 1992. *Green delusions: an environmentalist critique of
radical environmentalism*. Durham: Duke University Press.

McCray, W. P., 2005. "Will Small be Beautiful? Making Policies
for our Nanotech Future", *History and Technology*, 21(2): 177–
203.

Meadows, D. H., Meadows, D. L., Randers, J. and Beherens III,
W.W., 1972. *The limits to growth*. New York: New American
Library.

Medawar, P. B., 1967. *The art of the soluble*. London: Methuen

Merton, R. K., 1973 (1942). "The Normative Structure of Science",
in *The Sociology of Science: Theoretical and Empirical Investiga-
tions*. Chicago: University of Chicago Press.

Merton, R. K., 1968. Science and Democratic Social Structure, in
Social Theory and Social Structure: 604–615. New York: Free
Press.

Mitroff, I. I., 1974. "Norms and Counter-Norms in a Select Group
of the Apollo Moon Scientists: A Case Study of the Ambivalence of Scientists", *American Sociological Review*, 39(4): 579–
595.

Nature, (1990). 344.

Nordman, A., 2004. "New spaces for old cosmologies," *IEEE
Technology and Society Magazine*, Winter: 48-54.

Palmisano, S., 2008. "A smarter planet: The Next Leadership Agenda". Council on Foreign Relations, New York, 8 November.

Palmisano, S., 2010. "Welcome to the decade of smart". Royal Institute of International Affairs Chatham House, London, 12 January.

Palmisano, 2013. "How to compete in the era of smart". http://www.ibm.com/smarterplanet/global/files/us__en_us__overview__win_in_the_era_of_smart_op_ad_03_2013.pdf

Perrow, C., 1984. *Normal accidents: living with high risks technologies*. New York: Basic Books. Republished in 1999 with additions, Princeton University Press.

Polanyi, M., 1962. "The republic of science", *Minerva* 1: 54–73.

Popper, K., 1935. *The logic of scientific discovery*. Vienna: Verlag von Julius Springer.

Ravetz, J. R., 1991. "Ideological commitments in the philosophy of science", in Mynevar, G. (ed.), *Beyond Reason: Essays on the philosophy of Paul Feyerabend*: 355-377. Boston Studies in the Philosophy of Science, v. 132.

Ravetz, J. and Funtowicz, S., 2015. "Science, New Forms of", in Wright, J. D., (ed.), *International Encyclopedia of the Social and Behavioral Sciences*, 2nd edition, Vol. 21: 248–254. Oxford: Elsevier.

Rockström, J., Steffen, W., Noone, N., Persson, Å., Chapin, F.S. III, Lambin, E. F., Lenton, T. M., Scheffer, M., Folke, C., Schellnhuber, H. J., Nykvist, B., de Wit, C. A., Hughes, T., van der Leeuw, S., Rodhe, H., Sörlin, S., Snyder, P.K., Costanza, R., Svedin, U., Falkenmark, M., Karlberg, L., Corell, R. W., Fabry, V.J., Hansen, J., Walzer, B., Liverman, D., Richardson, K., Crutzen, P., and Foley, J. A., 2009. "A safe operating space for humanity", *Nature*, 461: 472-475.

Ruckelshaus, W. D., 1985. "Risk, science, and democracy", *Issues in Science and Technology*, 1(3): 19-38

Sarewitz, D., 2015. "Science can't solve it". *Nature*, 552: 413-414. 25 June.

Thompson, C., 2010. "What is I.B.M.'s Watson?", *New York Times*, 16 June.

United Nations, 1992. "Report of the United Nations Conference on Environment and Development" (Rio de Janeiro, 3-14 June 1992). Annex I. Rio Declaration on Environment and

Development.
http://www.un.org/documents/ga/conf151/aconf15126-
1annex1.htm.

Vermesan, O., Friess, P., Guillemin, P., Gusmeroli, S.,
Sundmaeker, H., Bassi, A., Soler Jubert, I., Mazura, M.,
Harrison, M., Eisenhauer, M. and Doody, P., 2011. "Internet
of Things Strategic Roadmap", IERC – European Research
Cluster on the Internet of Things.

Watson, J., 1968. *The double helix: a personal account of the discovery
of DNA*. New York: Atheneum.

Weinberg, A. M., 1967. *Reflections on Big Science*. Cambridge: MIT
Press.

Weinberg, A.M., 1972. "Science and trans-science", *Minerva*, 10:
209-222.

Weinberg, A.M., 1994.*The First Nuclear Era: The Life and Times of a
Technological Fixer*. New York: AIP Press.

Wildavsky, A., 1979. *Speaking truth to power*. Boston: Little,
Brown and Co.

Wilsdon, J., 2014. "Evidence-based Union? A new alliance for
science advice in Europe", *The Guardian*.
http://www.theguardian.com/science/political-sci-
ence/2014/jun/23/evidence-based-union-a-new-alliance-
for-science-advice-in-europe

4

INSTITUTIONS ON THE VERGE: WORKING AT THE SCIENCE POLICY INTERFACE[1]

Ângela Guimarães Pereira and Andrea Saltelli[2]

Introduction

In this chapter we set out to investigate the plausibility and the implications of two main hypotheses: (1) that the Joint Research Centre (JRC) of the European Commission (EC) is a "boundary institution" (Guston, 2001) at the intersection of scientific and policy spheres and as such is endowed with a unique role within the overarching structure of the EC; and (2) that this role is currently being challenged by an environment of decreasing trust in sci-

---

[1] The present chapter is based on and adapted from the following report published by the European Commission: Guimarães Pereira, Â. and Saltelli, A., 2014. "Of styles and methods. A quest for JRC's identity at times of change", JRC Technical Report EUR 26838 EN.
[2] The opinions expressed in the present work are those of the authors and cannot be taken to represent the views of the Joint Research Centre of the European Commission.

ence and by concurrent crises in the practice and governance of science.

We argue that the JRC operates at the intersection of overlapping political, societal and business spheres and that in this recent context the original meanings of its constitutional principles of authority, neutrality and independence can no longer be taken for granted. We offer a few suggestions on how to approach this challenge, with a view to transforming it into an opportunity for the organization.

We first try to understand what models of science and policy have historically underpinned the work of the JRC; next we proceed to identify recent developments, tensions and public debates affecting the practice and governance of science and science-based policy advice. Finally, we contend that an understanding of the JRC as a boundary institution would do justice to its uniqueness within the EC and the European Union (EU).

1. What the history of the JRC can tell us

We begin our case study of the JRC with a brief history of the organization, its origins, mandates and relationship to other EU institutions, exploring how the JRC has responded to evolving models of science and policy over a period of about 20 years. This account is drawn from an institutional brochure from 2007 celebrating the JRC's 50th anniversary (European Commission, 2007a) and from a 'Pictorial History' of the organization from 2009.

The JRC is today a Directorate-General (DG) of the EC, acting as a reference centre for research-based policy support in the EU. Historically, the JRC developed first as a joint nuclear research centre, following the signature in 1957 of the European Atomic Energy Community (EURATOM) treaty by six European countries. In fact, as "the nuclear industry started to expand at an unprecedented rate, national authorities in many European countries considered it critical to be able to develop nuclear knowledge: for example, neutron data were urgently needed for reactor design, waste management and reactor safety calculations" (European Commission, 2009: 8).

Hence, in 1959 the Ispra site of the JRC was inaugurated, with the Ispra-1 nuclear reactor completed within the first year of construction. The site became part of the then "European Community" in 1961. Throughout the 1970s the scope of the JRC was diversified in response to the diminishing urgency of nuclear research, the emergence of new themes worthy of European-level research, and the need for wider collaboration and greater coordination in European research. This led to JRC programmes on renewable energy, informatics and materials research, which were eventually formalized in 1973 with the introduction of a multi-annual research work programme overseen by a committee of experts from the Member States (MS).

During the 1980s, a major focus was on establishing research partnerships with industry in order to increase European competitiveness, with the launch of industry-related research programmes and collaborations (European Commission, 2009). The JRC was increasingly involved in collaborative projects and cooperation with national research bodies across the EU.

In the 1990s important research programmes at the JRC focused on public health, safety and security. This move into entirely new fields reflected the challenges and developments of the time. At the end of the 1990s, food scares such as BSE (bovine spongiform encephalopathy, commonly known as 'mad cow disease') and dioxin contamination led to the creation of the Directorate General Health and Consumer Protection (today's DG SANCO), separating the domain of food safety from that of industry and the environment; at the JRC the Institute for the Health and Consumer Protection (IHCP) was created in response to a number of consumer-related 'fiches' which are still relevant today. Another extension of the JRC was the establishment of the Institute for Prospective Technological Studies (IPTS), which responded to "the need to address new policy challenges involving both socio-economic and a scientific or technological dimension" (European Commission, 2009). The JRC followed the trend for restructuring across Europe, by merging institutes, re-naming some, and broadening the organization's research portfolio. Throughout these developments the JRC continued to assert its mission to provide impartial technical advice on relevant policy fiches (see Box 1).

During the 2000s a number of what became known as "Community Reference Laboratories" (CRLs) were established in various fields: feed additives, heavy metals, my-cotoxins, polycyclic aromatic hydrocarbons, genetically modified (GM) food and feed, and food contact materials.

In the mid-2000s, reflexive and anticipatory activities were given a more prominent place. The role of social sciences — though marginal — was promoted, especially in the field of quantitative economic analysis and techno-economic foresight, as well as via input to impact assessment studies. Employment, education, taxation, the single market and financial stability became part of the JRC's remit.

Box 1 compares two mission statements of the JRC a decade apart. Whilst the overall mission remained un-changed, and independence was maintained as a core value, some interesting changes are evident. For example, the 'customer-driven' approach of the early 2000s has been substituted by a principal set of customers: the policy Directorates General (DGs) of the EC. The JRC still collab-orates with Member States but the focus is now on coop-eration with its institutional partners.

Box 1. The changing mission of the JRC over the last decade

2002: The mission of the JRC is to provide **customer-driven scientific and technical support for the conception, development, implementation and monitoring of EU policies**. As a service of the European commission, the JRC functions **as a reference centre** of science and technology for the Union. Close to the policy making process, **it serves the common interest of Member States**, while being **independent of social interests whether private or national**.	**2013**: As the Commission's in-house science service, the Joint Research Centre's mission is to provide **EU policies with independent, evidence-based scientific and technical support throughout the whole policy cycle. Working in close cooperation with policy Directorates-General, the JRC addresses key societal challenges** while stimulating **innovation** through developing new methods, tools and standards, and sharing its know-how with the Member States, the scientific community and international partners.

Over the course of its history the JRC has addressed pressing societal and policy issues in a manner consistent with the predominant scientific narratives of the time, which included the assumption of authority, control, predictive power, independence, objectivity and the neutrality of science and scientific advice to policy. The evolution of the JRC's mission is a story of the gradual adoption of the role of independent scientific adviser, as well as of consensus builder, via its extensive work on standardization, reference methods, tools and laboratories.

1.1. Mapping the history of the JRC onto models of science and policy

Several models have been proposed to describe the relationship between science and decision making in policy processes. Funtowicz (2006) offers an evolutionary perspective on the approach to using science in policy making, starting from the assumption of scientific perfection and human perfectibility (the 'modern model'), and progressively incorporating elements of doubt and reflexivity, giving the following taxonomy:

- *Modern model.* This exemplifies the Cartesian vision of the limitless moral progress of humanity and its control over the environment. A defining characteristic of modernity is the relation between science and power, with the former offering legitimacy to the latter.

- *Precautionary model.* Precaution is introduced as a normative element. The model focuses on uncertain and inconclusive information. It arises from the discovery that the scientific facts are neither fully certain in themselves, nor conclusive for policy. 'Unintended consequences' are liable to follow from policies that have supposedly been rigorously designed.

- *Framing model.* Stakeholders' perspectives are introduced. The model arises from the recognition that in the absence of conclusive facts, scientific information becomes one among many inputs to a policy process, functioning as evidence rather than providing a logical demonstration or conclusive proof. Stakeholders' perspectives and values become relevant; even the choice of the scientific discipline to which 'the problem' belongs becomes a prior policy decision.

- *Demarcation model.* The focus is on protecting science and scientists from political interference that could threaten their integrity. This model is concerned with the possibility of the abuse of science; scientific information and advice that are used in the policy process are created by people working as employees in institutions with their own agendas. It recognises that 'scientific' information and advice cannot be guaranteed to be objective and neutral, as science can be abused when used as evidence in the policy process (see Chapron, 2014; Goldacre, 2012).

- *Extended participation.* The ideal of rigorous scientific demonstration is succeeded by that of open public dialogue, in which citizens become both critics and creators in the knowledge production process as part of an extended peer community. The model acknowledges the difficulties of defending a monopoly of accredited expertise on the provision of scientific information and advice. 'Science' (understood as the activity of technical experts) is included as one source of the relevant knowledge which is brought to bear as evidence.

Considering its history and the mission statements of the JRC, it seems that the institution has remained firmly attached to the core precepts of the 'modern model', *i.e.* "the experts' (desire for) truth speaking to the politicians' (need for) power" (Funtowicz and Strand, 2007) but also somehow to the 'demarcation model'. The JRC barely engages with the public, neither in the definition of its mission and portfolio nor in its actual operation. At the EC, institutionalized forms of public engagement are implemented by policy DGs through mechanisms such as the portal "Your Voice in Europe"[3], and through the inclusion of civil society organizations in committees and task forces, which are part of the EC's impact assessment practices. Although these activities have some value for the policy cycle, they remain limited to their consultative function and reach very small numbers of European citizens. Recently, the "European Citizens' initiative" has provided a further mechanism for citizen involvement in European policy[4]; projects such as "Voices"[5] have been showcasing the value of other types of participatory practices.

---

[3] See http://ec.europa.eu/yourvoice/

[4] See http://ec.europa.eu/citizens-initiative/

[5] See http://www.voicesforinnovation.eu/

1.2. The Uniqueness of the JRC

In an earlier working version of this chapter (Guimarães Pereira and Saltelli, 2014) we have compared the JRC to two other complex organizations: the Organisation for Economic Co-operation and Development (OECD) and a national research institute with a strong international presence, the Fraunhofer-Gesellschaft — Europe's largest application-oriented research organisation. To summarize, the JRC is not an institution that *promotes* policies like the OECD; it is neither a research and educational institution like a university, nor an institution that conducts research "to benefit private and public enterprise" as do the individual Fraunhofer Institutes. The JRC is a Directorate-General of the European Commission, but unlike other DGs it has some characteristics of a "boundary organization" (Guston, 2001). These types of organizations meet three criteria: "first, they provide the opportunity and sometimes the incentives for the creation and use of boundary objects[6] and standardized packages; second, they involve the participation of actors from both sides of the boundary, as well as professionals who serve a mediating role; third, they exist at the frontier of the two relatively different social worlds of politics and science, but they have distinct lines of accountability to each" (Guston, *ibid.*: 400-401).

We would argue that to some extent the JRC fits these criteria, given its role as facilitator or mediator of input from science to policy, from research institutes to DGs, dealing with different bodies of knowledge.

---

[6] *I.e.* information or knowledge that is used in different ways by different communities and networks. This concept was introduced by Star and Griesemer (1989).

2. Current challenges for science advice to policy

In order to put the role of the JRC in perspective, we will summarize in this short section a wealth of scholarship on the conflictual relation between science and policy. As we cannot do justice here to a century of epistemological dispute (see an attempt in Sarewitz, 2000; also Chapter 2 and Chapter 3, this volume), we will rather point to current challenges that affect science and technology, and therefore also the provision of scientific advice in support of public policy, with which the JRC is tasked.

The 18th-century vision of science as the future solution to all practical and social problems has been the subject of considerable debate and critique. Thinkers such as Husserl, Toulmin, Lyotard, Feyerabend, Lakatos and many others have questioned the putative duty and capacity of science to generate truth and resolve disputes. An early critique of the capacity of post-World War II science to tackle 'practical' (*i.e.,* 'policy') problems was offered by Ravetz (1971), in whose view the characteristics, problems and ethos of science have evolved since the 17th century. Problems arise when:

> *"[...] an immature or ineffective field is enlisted in the work of resolution of some practical problem. In such an uncontrolled and perhaps uncontrollable context, where facts are few and political passions many, the relevant immature field functions to a great extent as a 'folk-science'. This is a body of accepted knowledge whose function is not to provide the basis for further advance, but to offer comfort and reassurance to some body of believers. (Ravetz, 1971: 366)*

Science today suffers from crises of legitimacy (Lyotard, 1979), creativity (Le Fanu, 2010; Strumsky *et al.*, 2010) and quality (Ioannidis, 2005; Mirowski, 2011). Various attempts to characterize the state of affairs have been made, offering a number of alternative framings. "Post-normal science" (PNS) is a concept developed by Funtowicz and Ravetz (1991, 1992, 1993), which proposes a methodology of inquiry that is appropriate when "facts are uncertain, values in dispute, stakes high and decisions urgent" (Funtowicz and Ravetz, 1992: 251–273). "Mode 2 Science", coined in 1994 by Gibbons *et al.*, refers to a mode of production of scientific knowledge that is context-driven, problem-focused and interdisciplinary — see Carrozza (2015) for a recent discussion of these epistemologies.

2.1. Challenges from within...

The challenges described below affect the quality, integrity, authority and legitimacy of science advice to policy.

2.1.1. A challenge to trust

In his 2009 inaugural address, President Barack Obama promised to "restore science to its rightful place" in U.S. society. An interdisciplinary workshop organized at Arizona State University in 2010[7] set out to reflect on the meaning of such a political "gift". What is that place, it was asked, and how do we find it in an ever more complex, uncertain, and politically, socially and culturally diverse world? In late 2014, some felt confident enough to posit that the European Union's future lay in science (Malik, 2014), but this certainty was not universally shared.

At stake is the privileged role of scientific knowledge in legitimizing a common authority in secular, pluralist societies. Shapin and Schaffer have argued that "solutions to the problem of knowledge are solutions to the problem of social order" (1985: 332). A similar point had been made earlier by Jean-François Lyotard in his 1979 work *La Condition postmoderne. Rapport sur le savoir*. He adds that knowledge (identified with science) undergoes a process

---

[7] See http://cspo.events.asu.edu/

of 'delegitimization' when it becomes an industrialized commodity, as opposed to an instrument of emancipation and betterment of human beings (the German concept of *Bildung*, the ideal of traditional university education). He situates this process in the context of the end of the 'grand narratives' (of progress, Enlightenment, emancipation, *etc.*) that sustain modern Western culture.

More recently, based on the analysis of a number of case studies, Braun and Kroop have suggested that, "the expectation that scientific expertise will provide reliable, objective, true knowledge and thereby close down policy controversies is gone" (2014: 776).

The framing of issues in narrowly scientific terms can amount to what is described as 'Type 3 error' — that is, the error of trying to answer the wrong question. If the question is wrong, the evidence gathered is irrelevant. An issue may be framed as one of the 'risk' of a technology, when the concerns of citizens revolve rather around whose technology is being adopted, why, and who controls it. Scientific framings do not necessarily resolve socio-political controversies although they may appear relevant and convenient to some of the interested parties. Using a number of examples, from climate change to genetically modified organisms (GMOs) and nuclear waste disposal, Sarewitz (2004) has described the exacerbation of controversy through *scientized* framings that misrepresent the actual divisive issues.

It is a common practice for researchers to seek (through at times paternalistic approaches) a model — be it behavioural, psychological or cultural — to explain public dissent (Winner, 1986; Wynne, 1993). This can also justify overt attempts to manipulate public and media opinion to overcome dissent and disengagement. In the experience of the authors, when stakes are high, not all voices are paid equal attention (for example, in relation to GMOs and the Internet of Things); the representation of the issues offered by the media, business and sometimes governments are often impoverished accounts of the full range of perspectives. An example is what Bittman (2013) describes as an exercise of misdirection *vis-à-vis* organic foods. The study (see Smith-Sprangler *et al.*, 2012) focuses on a trivial aspect of the comparative advantages of organic versus conventional food, namely the poorly defined 'nutritional' value of organic food, when that is not in fact the primary reason that people consume organic food. Another example is the case of conflicting representations of GMO foods, described in Saltelli and Giampietro (this volume).

Even issues which once seemed to invite a linear treatment, from scientific advice to corrective policy, have today become 'wicked problems' (Rittel and Webber, 1973), meaning that they are now recognised to be deeply entangled in a web of barely separable facts, interests and values (see also Box 2). This is the case of GMOs, climate change, pesticides and bees, shale gas 'fracking', *etc.*

Box 2. On models

"Overreliance on model-generated crisp numbers and targets recently hit the headlines again in relation to the 90% ratio of public debt to gross domestic product stipulated by Harvard professors Kenneth Rogoff and Carmen Reinhart. Debt ratios above the threshold were considered by these authors as unsafe for a country, but a later reanalysis by researchers from the Univ. of Massachusetts at Amherst disproved this finding by tracing it to a coding error in the authors' original work. Critically, the error was corrected too late and much of the damage could not be undone, as the original model results kept austerity-minded economists trading blows with their anti-austerity counterparts on the merits and demerits of balanced budgets and austerity policies, a battle that dominated the financial press for months, and which was in no way defused by the repudiation of the Rogoff-Reihnart results." (in Saltelli *et al.*, 2013; see also Saltelli and Funtowicz, 2014)

We live in an era in which the media openly challenge public trust in science (Monbiot, 2013); norms associated with the scientific endeavour come under concerned scrutiny (Jasanoff, 2013); published results of laboratory experiments cannot be trusted (Sanderson, 2013); and software is on offer to help identify suspected work in preclinical cancer papers published in top tier journals (Begley, 2013).

The scientific community has long portrayed its endeavour as self-regulating, inasmuch as it has a higher ethical commitment to truth-telling than other sectors of society as a whole. Yet the tone and intractability of present controversies suggest that society may now be less willing to accept such claims than in the past. Something worth recalling is that the scientific enterprise depends on a certain fundamental *ethos*. In the words of Ravetz (1971: 22):

> *Two separate factors are necessary for the achievement of worthwhile scientific results: a community of scholars with shared knowledge of the standards of quality appropriate for their work and a shared commitment to enforce those standards by the informal sanctions the community possesses; and individuals whose personal integrity sets standards at least high as those required by their community. If either of these conditions is lacking (…) then bad work will be produced. (…) Any view of science which fails to recognize the special conditions necessary for the maintenance of morale in science is bound to make disastrous blunders in the planning of science.*

Many scholars have asserted that the authority of science resides above all in its accommodation of dissent and openness to criticism. The classic work of Popper was based on the assumption that criticism is the essence of science — see his *Conjectures and Refutations* (1963). But it is also assumed that there must be a consensus that ends the argument in order for science to be authoritative (Hulme, 2013). Sarewitz has criticized this belief, arguing that "[s]cience would provide better value to politics if it articulated the broadest set of plausible interpretations, options and perspectives, imagined by the best experts, rather than forcing convergence to an allegedly unified voice" (2011: 7).

Another aspect of this discussion on trust relates to ethics. The application of ethics in the realm of science and technology has largely been in the hands of professional communities. Yet there have been multiple grievous ethical failures in science: for example, the use of humans in chemical research on poisonous gas during World War I ("the chemists' war", Ravetz, 1971: 38); the role of statistical science in underpinning eugenics (Hacking, 1990); the use of humans in experiments in World War II (U.S. Holocaust Memorial, 2015); and the systematic implication of science (better described as technoscience) in the generation of ever more and greater environmental challenges. The recognition of the fallibility of the scientific community has given ethics an increasingly important role in addressing questions of values and conscience in relation to technoscientific developments.

Today, the existing institutional arrangements and ethical frameworks are being challenged by new ways of doing science, including emerging alternative movements and spaces such as 'garage science', 'maker spaces', 'do-it-yourself' movements, (such as DIYbio or Do-it-Yourself Biology — see Delgado, 2013), hacker spaces and also by emerging paradigms such as the "Open Source Everything" paradigm and crowdfunded research. With an enlarged set of actors bringing their norms and values to technoscientific knowledge production, not only have the *loci* of ethics been re-distributed and extended (Toulmin in Lifson, 1997; Tallacchini, 2009, 2015), they have also been systematically interrogated by citizens, as well as by traditional and new media. Digital culture has accelerated this process by enabling diverse epistemic networks and more tools to come into existence.

2.1.2. A challenge to quality

Uncertainty is at the heart of debate about quality in science; it is standard practice by stakeholders to either minimize or maximize the perception of uncertainty in order to, respectively, support or dispute a particular action or policy. Famous cases of the fabrication of uncertainty were the denial by tobacco companies of the deleterious health effects of smoking (Oreskes and Conway, 2010), and the battles between industry and regulators over the USA's data quality act, in which industry fought to amplify uncertainty in order to prevent regulators from imposing more stringent standards (Michaels, 2005). As discussed earlier, scientists may themselves encumber the public debate with a supplementary dose of conflict and animosity, making controversies less amenable to a solution (Sarewitz, 2004).

"Another busy week at Retraction Watch"

The rise in the number of retractions of scientific papers has been raising concern for several years (see, for example, Van Noorden, 2011, and the website Retraction Watch[8]), mostly because many, if not most, of the retractions that hit the headlines involve dramatic cases of misconduct, while few appear to be due to honest error. This situation, in combination with cases of the non-verifiability of experiments and data, poses challenges for the maintenance of the quality of scientific publications — a problem which is even more acute for policy-relevant science. An example of how serious this may become is the recently proposed legislation in the USA, "To prohibit the Environmental Protection Agency from proposing, finalizing, or disseminating regulations or assessments based upon science that is not transparent or reproducible"[9].

---

[8] See http://retractionwatch.com

[9] See https://beta.congress.gov/bill/113th-congress/house-bill/4012

According to Mirowski (1991, 2013), one of the consequences of having adopted neoliberal policies and a neoclassical stance in economics since the 1980s has been a massive privatization and commoditization of research, with a serious effect on the system of self-governance of science and on the quality of scientific output (see also Saltelli, Ravetz and Funtowicz, this volume). The link between commoditized research and the reproducibility debacle led the magazine *The Economist* to talk of "shoddy research" (2013) and "sloppy researchers" (2014).

According to Masood (in Guimarães Pereira, 2012: 23), "Scientists, policymakers and publishers regard peer review as the gold standard in science. But how true is this in a world where the very idea of expertise and authority is open to question? Does conventional peer review make sense in a world in which anyone with a mobile phone, a WiFi connection and a Twitter account is both reader and reviewer?" Indeed, the rise of algorithms allows for both fraud and detection of fraud in publications; the publisher Springer has recently developed software ('SciDetect') that detects algorithmically generated papers (Springer, 2015).

The history of quality assurance in open-source software development (and in shareware and freeware, for that matter) is relevant for the world of extended peer review. Over the years the 'open source communities' have come to follow the Open Source Software Development (OSSD) standard, an accepted quality process.

The rise of digital publishing needs to be accompanied by new, collectively agreed methods of quality assurance. These new strategies of quality assurance will not be just about the publications, but also about the research framings, agendas, questions, and assumptions needed to address societal concerns.

From an optimistic standpoint, it could be argued that digital culture has at last permitted the "extended peer review" and strengthened quality assurance systems advocated by Funtowicz and Ravetz (1990) to gain a foothold. The extended peer review model improves quality assurance by taking into account different types of knowledge and involving an extended community of social actors.

In addition, policy-makers themselves are calling for better quality assurance mechanisms and standards in policy-relevant science; Ian Boyd, science adviser to the Department for Environment, Food and Rural Affairs of the UK government (DEFRA) has expressed "concern about unreliability in scientific literature" and "systematic bias in research", suggesting that an auditing process would help policy-makers navigate research bias (Boyd, 2013).

The issue of quality has become ever more important in the DIY era. Digital culture and phenomena like DIY science (Nascimento *et al.*, 2014), citizen science (Irwin, 1995; Bonney, 1996) and the "Open Source Everything" paradigm (Steele, 2014), all involving a broad community of actors in the production and preservation of knowledge, certainly have effects on the knowledge production process and on the assessment and governance of scientific institutions. The broader and deeper involvement of society in the scientific enterprise is largely an outcome of voluntary individual and community action. This grassroots engagement takes various forms and modes — *e.g.* self- and *sousveillance*[10], crowd-funded initiatives, hacker spaces, maker spaces, *fablabs*, community ICT-based research, *etc*. This extension of participation

---

[10] 'Sousveillance' "is the recording of an activity by a participant in the activity, typically by way of small wearable or portable personal technologies" (see Wikipedia). The term is owed to Steve Mann.

calls for reflection on the nature of the knowledge produced, and on the criteria and processes for assuring its quality and integrity.

2.1.3. *Challenges to legitimacy and democracy*

In Europe, the BSE 'scandal' in the United Kingdom from the mid-1980s to mid-1990s is often cited as having been pivotal in the change of direction in the relation between science and policy making. A key moment was the publication of the House of Lords report on Science and Society (House of Lords, 2000), followed a year later by the European Commission's "Science and Society Action Plan" (European Commission, 2002), and then by the EU 5th Framework Programme's "Raising Awareness of Science and Technology" activity in the late 1990s.

The BSE crisis was perhaps instrumental in calling into question the so-called 'deficit model' (which was the great inspiration for the Public Understanding of Science — 'PUS' — movement[11]). The deficit model — which explains public opposition to new technologies as the result of a generally poor understanding of science — is, in fact, still alive, although it has been challenged for a long time. Dan Kahan, a theorist of cultural cognition, has recently argued (2015) that opposition to climate science is not due to lack of scientific knowledge but rather the opposite; more polarized opinions are actually found among better in-

---

[11] Public understanding of science can be described as a movement that aimed at redressing a 'deficit' in the public's knowledge about science. This movement sought to engage scientists from the 1980s onwards in one-way communication of scientific processes, in which scientists were invited to educate the scientifically illiterate publics about scientific knowledge as a means of dealing with public opposition to science and technology. See, for example, Miller (2001).

formed people, the debate being driven by powerful normative and cultural stances.

Yet, public views of science worsened throughout the 1990s, and a new discourse of 'science and society', dialogue and engagement emerged. It was also during the 1990s that discussions of ethical issues in domains such as environment and medicine came out of the cave of the professional community and reached wider publics (Toulmin, cited in Lifson, 1997).

Still, as Jasanoff (2005) notes, in some important cases upstream efforts to identify risks and explore ethical dilemmas were led by the scientific community itself. A remarkable example of government initiative was the UK state-sponsored debate on genetically modified crops, "The GM Nation?", which is often seen as a reaction to the BSE crisis (Gaskel *et al.*, 2003). In the EU, public engagement and ethics were at the heart of the "Responsible Research and Innovation" (RRI) initiative. It is instructive to attend to how the names of EC research programmes addressing public interfaces evolved: from "Science *and* Society" to "Science *in* Society", followed by "Science *with* Society", and most recently, in "Horizon 2020", "Science *in and with* Society".

Ulrich Beck (1992) described as "reflexive modernity" the state in which growing bodies of knowledge are accessible to growing numbers of individuals with increased agency that enables them to intervene in the world. Several authors have anticipated this state of deeper involvement of non-experts in scientific treatment of societal matters; for example, Funtowicz and Ravetz (1990) call for reflexivity through "extended peer communities" and "extended facts", concepts which are at the core of the model of post-normal science; Callon *et al.* (2001) describe it as the "public dialogue and participation model", whilst Jasanoff (2005) describes this configuration with the concept of "civic epistemologies", *i.e.* "the broader array of activities, social processes, informal practices, and institutionalized procedures by which people collect, aggregate, validate, and wield claims to knowledge about nature and society in public and policy settings" (after Miller, 2005: 410).

These concepts represent an explicit rejection of the deficit model and the idea that the public is incapable of understanding or reflecting on scientific issues; they attest rather to efforts to democratize science. The past decade has witnessed crowd-funded radioactivity measurements in the aftermath of Fukushima (McNeill, 2014), the Quantified Self movement that deals with *self-veillance* of health issues and a general drive toward commons-based "peer production" of knowledge (Benkler and Nissembaun, 2006).

In his recent book, *The Open-Source Everything Manifesto – transparency, truth and trust,* the former Central Intelligence Agency case officer Robert D. Steele argues that the open-source-everything paradigm is the condition *sine qua non* for restoring public trust in the wake of deep corruption and secrecy that have enabled widespread fraud across private and public institutions; he furthermore advocates for decision making that is based on collective and bottom-up action to address major world crises (Steele, 2014).

In short, challenges to democracy and the legitimacy of science cannot be addressed without a democratization of the interface between science and its publics.

2.2. Challenges from without

Science is experiencing a crisis of quality, trust and legitimacy which affects both its practice and its ethos. Scientific advice is dependent upon the context in which science develops. In other words, science and scientific advice are co-produced and interwoven. How will the crises in science affect the 'evidence-based policy' paradigm?

2.2.1. From evidence-based policy to policy-based evidence?

The assumptions and weaknesses of the evidence-based policy paradigm and the demarcation model that underpins it are discussed at length elsewhere in this volume (Chapter 2, Chapter 3).

However, there is an even more problematic aspect to this narrative: the danger of evidence-based policy turning into policy-based evidence, not necessarily because of pressure from lobbyists or opportunism among the responsible policy makers, but often because of constraints attendant on short policy cycles. It is in these settings that the standards for quality must be set high, both for the internal institutional peer review process as well as for the consultation of stakeholders. This has implications for ensuring integrity and trust in institutions like the JRC, as we shall discuss later.

2.2.2. Anticipatory culture

A 'consequentialist culture' seems to subsist in the governance of science and technology — that is, we tend to look for 'consequences' (impacts, risks, *etc.*) rather than for the meanings and the narratives that sustain science and technology. We allude here briefly to the extensive research on the culture of using cost–benefit analyses, risk assessment and other types of normative approaches as a basis for 'consequentialist governance'. This is a field in its own right: see, for example, Collingridge and Reeve (1986), Krimsky and Golding (1992), Perrow (1984), Funtowicz and Ravetz (1990), Jasanoff (2010), Taleb (2007, 2012) and the European Commission (2007b).

Suffice it to say here that there are cases where cost–benefit analyses are pushed too far, quantifying the unquantifiable, or where the acceptability of a new technology is expediently reframed as an issue of risk. In the words of Langdon Winner (1986):

> [T]he risk debate is one that certain kinds of social interests can expect to lose by the very act of entering. [...] Fortunately, many issues talked about as risks can be legitimately described in other ways. Confronted with any cases of past, present, or obvious future harm, it is possible to discuss that harm directly without pretending that you are playing craps. A toxic waste disposal site placed in your neighborhood need not be defined as a risk; it might appropriately be defined as a problem of toxic waste. Air polluted by automobiles and industrial smokestacks need not be defined as a 'risk'; it might still be called by the old-fashioned name, 'pollution'. New Englanders who find acid rain falling on them are under no obligation to begin analyzing the 'risks of acid rain'; they might retain some Yankee stubbornness and confound the experts by talking about 'that destructive acid rain' and what's to be done about it. A treasured natural environment endangered by industrial activity need not be regarded as something at 'risk'; one might regard it more positively as an entity that ought to be preserved in its own right. (1986: 151)

We suggest that *anticipatory* governance is better suited to dealing with uncertainties and unknowns. Anticipation implies building the capacity to respond to unpredicted and unpredictable risks and indeterminacies (Guston, 2008) by cultivating participatory approaches to foresight that extend to the public the right to imagine possible futures and assess the visions produced.

In an earlier working version of the present chapter, Guimarães Pereira and Saltelli (2014) exemplified how this shift of culture would affect our reading of innovation and growth narratives. In no case should narratives be taken for granted or considered as panaceas. As a boundary organization, the JRC could help to create space to raise and discuss these kinds of questions. A positive indication of this potential new role for JRC was a series of work-shops it organized on issues relating to the quality of knowledge production for both mainstream and emerging science, and on the implications of these issues for poli-cy[12].

3. Futuring scientific advice to policy making: an 'Emperor's new clothes' model?

3.1. Cultures of advice

In this section we examine some discussions of science advice with relevance to a boundary institution such as the JRC.

There are different cultures of science advice, from the "heroic model" (Doubleday and Wilsdon, 2013) personified by Chief Scientific Advisers (CSAs) as in the UK and some other countries, to science academies, or models that rely on broader consultations with civil society, such as the Danish Board of Technology — see Box 3.

---

[12] See https://ec.europa.eu/jrc/en/event/workshop/new-narratives-innovation;
https://ec.europa.eu/jrc/en/event/workshop/do-it-yourself;
https://ec.europa.eu/jrc/en/event/conference/use-quantitative-information

Box 3.

> The Danish Board of Technology (DBT) (see http://www.tekno.dk) works at the interface between "public challenges, technology, knowledge, values and actions to be taken. The DBT [counsel] decision-makers about possibilities and consequences for citizens, environment and society and create platforms for participants to pool their knowledge, finding sustainable and interdisciplinary solutions. [It] works with developing dialogue based methods at a local, national and global level. The DBT furthermore implements projects at a national and an international level for the EU and globally in collaboration with the United Nations". The activities of the DBT span various technologies (e.g. biotechnology and ICT), and economic sectors (e.g. transport and agriculture. DBT methods are participatory by design, for example, Citizens' Summits, Citizens' Juries, Consensus Conferences.

As Pielke (2013) noted in his address to the UK Chief Scientific Adviser, 'science advice' is a misnomer as it implies a deficit model in which the 'advice' is artificially decontextualized from the political situation. Moreover, he argues, science advisers cannot "carry the authority of science as a counterbalance to the messiness of politics" (2013: 122). But, in what 'balanced' form, if that is possible, should an individual adviser act as 'spokesperson' for disparate, often dissenting scientific voices in fields prone to controversy?

Science advice needs to deal with the same kinds of tensions as the governance and practice of science described earlier. As Jasanoff (2013) notes, scientific advisers are bound by the same principles and discipline of scientific practices (*i.e.* "known facts, reliable methods, responsible professional codes, and the ultimate test of peer review": 62). For this reason, "science advisers are not inclined to introspection in situations where their work fails to persuade", even if "science advisers can offer at best educated guesses and reasoned judgments, not unvarnished truth" (*ibid.*: 62). Rather, she argues, it is often the case that such failures are attributed to factors that are external to the scientific endeavour, such as an ignorant and obstreperous public, or perverse media and corrupting "powerful corporate funders or other large interest groups". Yet, trust is a *conditio sine qua non* for the recognition of the role of science advisory systems; such systems and institutions need extended scrutiny, as any other democratic institution (Jasanoff , *ibid.*: 67).

The science of policy advice has stepped out of the shadows of academic and policy practice and is now closely scrutinized by the media. For example, the British newspaper *The Guardian* began in 2014 to maintain an informed commentary about science advice to policy, offering checklists of do's and don't's and principles for how scientific advisers should do their jobs[13].

This focus by the media on science advice to policy is interesting, to say the least; Petersen *et al.* (2010) investigated the transformation of scientific policy advice in relation to mass media, arguing that the increasing mediatization of science deeply affects the ways in which policy-makers utilize scientific expertise. Policy makers cannot afford to ignore scientific knowledge and controversies once these are published in the media, since the protagonists are often scientists known to and respected by the public. In one way or the other, we are witnessing a process of mass media expertise, which has altered the relationships between advisers and policy makers.

Hence, we argue that scientific advice is changing, not only because of the contextual issues enumerated in Section 2, but also because it is becoming a distributed endeavour, in which the various epistemic networks[14] employ diverse strategies to make their voices heard, in-

---

[13] See http://www.theguardian.com/science/science-policy

[14] "Epistemic network" is a concept developed by Kjetil Rommetveit and others within the EPINET project (http://www.epinet.no) on the basis of Haas's (1992) concept of "epistemic community".

cluding with the help of new media — posing a challenge to institutional practices.

3.2. Opportunities for science advice

In the preceding sections we discussed aspects of an internal and external crisis in science and knowledge production that affect not only scientific practice but also scientific advice.

In this section we would like to propose that scientific advice be seen as an opportunity for societal reflexivity. Reflexive practice in institutions like the European Commission is not only about examining the body of knowledge chosen to sustain particular claims or the legitimacy of various voices, but also and above all about the critical and comprehensive testing of prevailing narratives against a broader spectrum of worldviews.

Our thesis is that boundary institutions such as the JRC are in a unique position to organize reflexive thinking and action to address those crises. The JRC constitutes a model for science advice *per se*, being by design multi-disciplinary, multi-cultural and providing a space for dialogue between different values. The JRC must necessarily operate within the bounds of its constitution, but rather than pursuing its business unreflectively and according to received wisdom — what we call elsewhere "following the Cartesian dream" (Guimarães Pereira and Funtowicz, 2015) — it should explore different understandings of knowledge production, power and societal organization in order to respond to the ongoing crises.

We here propose a model that takes account of two important features of the JRC: on the one hand, its role in the policy cycle, and on the other, its natural intimacy with academia. We could call this 'the emperor's new clothes' model, after the story by Hans Christian Andersen, because this model may entail delivering an unwelcome message. A recent example of JRC work in this vein is Saltelli and Dragomirescu-Gaina (2014) and the workshop on "New Narratives for Innovation" mentioned above, which radically challenged some of the existing fantasies of innovation as an instrument to solve the present crisis in the EU (through the creation of "new and better jobs") and explicitly called attention to the crisis in science. These types of activities seem to be welcome and useful, as the testimony of workshop participants — many from EU institutions — illustrates: "to challenge some of our policy thinking, to test its robustness, and explicitly try to get some of the people like myself to listen to ideas that might not fit with the particular narrative that we are trying to sell [...] It is not always a comfortable thing to do, but we have to be open and acknowledge those weaknesses on our policy arguments" (workshop participant, quoted in Guimarães Pereira, Saltelli and Tarantola, 2015: 13).

Our model for the JRC emphasises a commitment to three complementary cultures:

1. *Quality*. This entails developing a practice of *extended peer review,* embracing emerging epistemologies, putting social sciences at the heart of the operation of the JRC. One objective would be to achieve social robustness of technoscientific proposals, as advocated in PNS and Mode 2 science, acknowledging that a broad spectrum of norms needs to be taken into account. A central issue for today's scientific enterprise is the link between quality and trust. The JRC could develop evaluation strategies and provide pedigrees for evidence-based policy. The JRC has a role to play not only in producing facts and figures, and in modelling, but also in the elicitation of the worldviews that sustain them, engaging all relevant social actors in a quality assurance process. Ultimately, quality assurance is about testing the credibility, fitness for purpose and

social robustness of what is offered as evidence to underpin policy making.

2. *Reflexivity.* As a boundary institution, the JRC is inherently reflexive, but space needs to be provided for its body of researchers to inquire critically and inquisitively into policy agendas and political imaginaries through processes of knowledge assessment. This reflexive model of scientific advice aims to challenge narratives based on business as usual and to test their relevance against social agendas and social values. Science is seen as the epitome of self-reflexivity, but the prevalence of the deficit model—which sees the public as ignorant, risk-adverse, unreflexive and disobedient—reveals that scientific institutions are often only weakly reflexive when it comes to the indirect objects of their work (Wynne, 1993). We suggest that this is a clear opportunity for the JRC, in as much as the available 'policy-based evidence' can be interrogated through knowledge assessment. A reflexive model calls for a great investment in anticipation, extended peer review, ethics, knowledge assessment, and upstream public engagement.

3. *Humility.* As discussed by Jasanoff (2003, 2007), "technologies of humility" imply developing a culture of *engagement*: firmly rejecting the deficit model and valuing dialogic governance, on the premise that in the face of different types of uncertainties and unknowns, the anticipation of impacts, determination of relevant facts and norms, the questions to be asked and the methods of enquiry to be employed are collective decisions not to be surrendered to powerful elites—not even to an elite of scientists.

This is in line with Funtowicz's (2006) model of extended participation as described in Section 1, in which citizens become both critics and creators in the knowledge production process. As a boundary institution, it would be a missed opportunity for the JRC not to pursue this line of engagement.

4. A final thought

We have considered the JRC as a 'boundary organization' that operates at the interface of science and policy in an era of decreasing trust and concurrent crises in the practice and governance of science. We have suggested that organizations in this situation should appeal to cultures of quality, reflexivity and humility. Our example, the JRC, as the "in-house science service" of an important international institution, is ideally positioned to consider new ways in which science can be deployed and to broaden and deepen its interfaces with society.

The main challenge faced by an organization such as the JRC in this process is that important actors of change—scientists themselves and their policy-making counterparts in the European Commission DGs—are at the same time engaged and committed to those existing practices and cultures which are most in need of change.

Acknowledgments

Useful suggestions and encouragements were received in an earlier version of this paper by Jerome R. Ravetz (Oxford University), and William Becker, Sjoerd Hardeman, Estefania Aguilar Moreno (JRC). Any errors remaining originate with the authors.

References

Beck, U., 1992. *Risk Society: Towards a New Modernity*. New Delhi: Sage.

Begley, C. G., 2013. "Reproducibility: Six red flags for suspect work", *Nature*, 497: 433–434.

Benkler, Y. and Nissenbaum, H., 2006. "Commons-based Peer Production and Virtue", *Journal of Political Philosophy*, 14(4): 394–419.

Bittman, M., 2012. "That Flawed Stanford Study", 2 October. http://opinionator.blogs.nytimes.com/2012/10/02/that-flawed-stanford-study/?_php=true&_type=blogs&_r=0

Bonney, R., 1996. "Citizen science: A Lab tradition", *Living Bird*, 15(4): 7-15.

Boyd, I., 2013. "A standard for policy-relevant science", *Nature Comment*, 501(12): 160.

Braun, K. and Kroop, C., 2014. "Beyond Speaking Truth? Institutional Responses to Uncertainty in Scientific Governance", *Science and Technology & Human Values*, 35(6): 771-782.

Callon, M., Lascoumes, P., Barthe, Y., 2001. *Agir dans un monde incertain. Essai sur la démocratie technique*. Paris: Le Seuil.

Carrozza, C., 2015. "Democratizing Expertise and Environmental Governance: Different Approaches to the Politics of Science and their Relevance for Policy Analysis", *Journal of Environmental Policy & Planning*, 17(1), 108-126.

Chapron, G., 2014. "Challenge the abuse of science in setting policy", *Nature*, 516 (7531).

Collingridge, D. and Reeve, C., 1986. *Science Speaks to Power: The Role of Experts in Policy Making*. London: Frances Pinter.

Delgado, A., 2013. "DYIbio: Making Things and making futures", *Futures*, 48: 65-73.

Doubleday R. and Wilsdon, J. (eds.), 2013. *Future Directions for Scientific Advice in Whitehall*. http://sro.sussex.ac.uk/47848/2/FDSAW_Wilsdon_Double day.pdf

Economist, 2013. "How science goes wrong", 19 October.

Economist, 2014. "Combating bad science Metaphysicians. Sloppy researchers beware. A new institute has you in its sights", 15 March.

European Commission, 2002. *Science and Society Action Plan*. Publications Office of the European Union, Luxembourg.

European Commission, 2007a. *Highlights of the JRC: 50 years in science*. Office for Official Publications of the European Communities, Luxembourg.

European Commission, 2007b. *Taking European Knowledge Society Seriously: Report of the Expert Group on Science and Governance to the Science, Economy and Society Directorate, Directorate-General for Research, European Commission*. Publications Office of the European Union, Luxembourg.

European Commission, 2009. *JRC Ispra: A 50 Year Pictorial History*, Office for Official Publications of the European Communities, Luxembourg, report number JRC49907. https://ec.europa.eu/jrc/sites/default/files/jrc_50_years_b rochure_en.pdf

Funtowicz, S., 2006. "What is Knowledge Assessment?", in Guimarães Pereira, Â., Guedes Vaz, S. and Tognetti, S. (eds.), *Interfaces between Science and Society*. 138-145. Sheffield: Greenleaf Publishers.

Funtowicz, S. and Ravetz, J., 1990. *Uncertainty and Quality in Science for Policy*. Dordrecht: Kluwer Academic Publishers.

Funtowicz, S. O. and Ravetz, J. R., 1991. "A New Scientific Methodology for Global Environmental Issues", in Costanza, R. (ed.) *Ecological Economics: The Science and Management of Sustainability*: 137–152. New York: Columbia University Press.

Funtowicz, S. O. and Ravetz, J. R., 1992. "Three types of risk assessment and the emergence of postnormal science", in Krimsky, S. and Golding, D. (eds.), *Social theories of risk*: 251–273. Westport, CT: Greenwood.

Funtowicz, S. O. and Ravetz, J. R., 1993. "Science for the postnormal age", *Futures*, 25 (7): 739–755.

Funtowicz, S. and Strand, R., 2007. "Models of Science and Policy", in Traavik, T. and Lim, L.C. (eds.), Biosafety First. Tapir Academic Publishers. http://genok.no/wp-content/uploads/2013/04/Chapter-16.pdf

Gaskell, G., Allum, N., Bauer, M. W., Jackson, J., Howard, S., and Lindsey, N., 2003. *Ambivalent GM nation? Public attitudes to biotechnology in the UK, 1991-2002*. London: London School of Economics and Political Science.

Gibbons, M., Limoges, C., Nowotny, H., Schwartzman, S., Scott, P., Trow, M., 1994. *The New Production of Knowledge: The Dynamics of Science and Research in Contemporary Societies*. London: Sage.

145

Goldacre, B., 2012. *Bad Pharma: How drug companies mislead doctors and harm patients*. London: Fourth Estate.

Guimarães Pereira, Â., 2012. *Science in a Digital Society*. European Commission, EUR 25201 EN – 2012.
http://publications.jrc.ec.europa.eu/repository/bitstream/J
RC68607/lbna25201enn.pdf.pdf

Guimarães Pereira, Â. and Funtowicz, S. (eds.), 2015. *Science, Philosophy and Sustainability: The End of the Cartesian Dream*. New York: Routledge.

Guimarães Pereira, Â. and Saltelli, A., 2014. "Of styles and methods. A quest for JRC's identity at times of change", JRC Technical Report EUR 26838 EN.
http://publications.jrc.ec.europa.eu/repository/bitstream/J
RC91736/straw-man%20final-tr_register.pdf

Guimarães Pereira, Â., Saltelli, A. and Tarantola, S., 2015. "New Narratives for Innovation", EUR report. European Commission, to appear.

Guston, D., 2001. "Boundary Organisations in Environmental Policy and Science: An Introduction", *Science, Technology, & Human Values*, 26(4): 399-408.

Guston, D., 2008. "Innovation Policy: not just a jumbo shrimp", *Nature*, 454: 940-941.

Haas, P. M., 1992. "Knowledge, Power, and International Policy Coordination", *International Organization*, 46(1): 1-35.

Hacking, I., 1990. *The Taming of Chance*. Cambridge: Cambridge University Press.

House of Lords, 2000. *Science and Society*. London: Her Majesty's Stationary Office.

Hulme, M., 2013. "Lessons from the IPCC: Do scientific assessments need to be consensual to be authoritative?" in Doubleday R. and Wilsdon, J. (eds.), 2013. *Future Directions for Scientific Advice in Whitehall*.
http://sro.sussex.ac.uk/47848/2/FDSAW_Wilsdon_Double
day.pdf

Ioannidis, J. P. A., 2005. "Why Most Published Research Findings Are False", *PLoS Medicine*, 2(8): 696-701.

Irwin, A., 1995. *Citizen science: a study of people, expertise, and sustainable development*. Psychology Press.

Jasanoff, S., 1996. "Beyond Epistemology: Relativism and Engagement in the Politics of Science", *Social Studies of Science*. 26(2): 393-418.

Jasanoff, S., 2003, "Technologies of Humility: Citizen Participation in Governing Science", *Minerva*, 41(3): 223-244.

Jasanoff, S., 2005. *Designs on nature: science and democracy in Europe and the United States*. Princeton: Princeton University Press.

Jasanoff, S., 2007. "Science & Politics. Technologies of humility", *Nature*, 450: 33.

Jasanoff, S., 2010. "Beyond calculation: a democratic response to risk", in Lakoff, A. (ed.), *Disaster and the politics of intervention*. New York: Columbia University Press.

Jasanoff, S., 2013. "The science of science advice", in Doubleday R. and Wilsdon, J. (eds.), 2013. *Future Directions for Scientific Advice in Whitehall*.
http://sro.sussex.ac.uk/47848/2/FDSAW_Wilsdon_Double day.pdf

Kahan, D. M., 2015. "Climate science communication and the measurement problem", *Advances in Political Psychology*, 36: 1-43.

Krimsky, S. and Golding, D. (eds.), 1992. *Social Theories of Risk*. Westport: Praeger.

Le Fanu, J., 2010. "Science's Dead End", *Prospect Magazine*, 21 July.
http://www.prospectmagazine.co.uk/features/sciences-dead-end

Lifson, A., 1997. "A Conversation with Stephen Toulmin", *Humanities*, 18(2).

Lyotard, J.-F., 1979. *La Condition postmoderne. Rapport sur le savoir*. Paris: Minuit.

Malik, X. (ed.), 2014. "The Future of Europe is Science; A report of the President's Science and Technology Advisory Council (STAC)". European Commission, October.

Marris, C., Wynne, B., Simmons, P. and Weldon, S., 2001. "Final Report of the PABE Research Project Funded by the Commission of European Communities", Contract number: FAIR CT98-3844 (DG12-SSMI) Dec. Lancaster: University of Lancaster.

McNeill, D., 2014. "Concerns over measurement of Fukushima fallout", *New York Times*, 16 March.
http://www.nytimes.com/2014/03/17/world/asia/concerns-over-measurement-of-fukushima-fallout.html?_r=0

Michaels, D., 2005. "Doubt is their product", *Scientific American*, 292: 6.

Miller, C., 2005. "New Civic Epistemologies of Quantification: Making Sense of Indicators of Local and Global Sustainability", *Science, Technology, Human Values*, 30(3): 403-432.

Miller, S., 2001. "Public understanding of science at the crossroads". *Public Understanding of Science*, 10(1): 115-120.

Mirowski, P., 1991. *More Heat than Light. Economics as Social Physics, Physics as Nature's Economics*. Cambridge: Cambridge University Press.

Mirowski, P., 2011. *Science-Mart: Privatizing American Science*. Harvard: Harvard University Press.

Mirowski, P., 2013. *Never Let a Serious Crisis Go to Waste: How Neoliberalism Survived the Financial Meltdown*. Brooklyn: Verso Books.

Monbiot, G., 2013. "Beware the rise of the government scientists turned lobbyists", *The Guardian*, 29 April.

Nascimento, S., Guimarães Pereira, Â., Ghezzi, A., 2014. *From Citizen Science to Do It Yourself Science*. Scientific and Policy report. European Commission: EUR 27095.
http://publications.jrc.ec.europa.eu/repository/bitstream/JRC93942/ldna27095enn.pdf

Oreskes, N. and Conway, E. M., 2010. *Merchants of Doubt: How a Handful of Scientists Obscured the Truth on Issues from Tobacco Smoke to Global Warming*. New York: Bloomsbury Press.

Perrow, C., 1984. *Normal Accidents: living with high risk technologies*. New York: Basic Books.

Petersen, I, Heinrichs, H. and Peters, H. P., 2010. "Mass-Mediated Expertise as Informal Policy Advice", *Science, Technology and Human Values*, 35(6): 865-887.

Pielke, R. Jr., 2013. "Letter from America: a memo to chief scientific adviser Sir Mark Walport", *The Guardian*, 15 April.

Popper, K., 1963. *Conjectures and Refutations: The Growth of Scientific Knowledge*. Routledge and Kegan Paul.

Ravetz, J. R., 1971. *Scientific Knowledge and Its Social Problems*. Oxford University Press.

Rittel, H. W. J. and Webber, M. M., 1973. "Dilemmas in a General Theory of Planning", *Policy Sciences*, 4: 155-169.

Saltelli, A. and Dragomirescu-Gaina, C., 2014. "New Narratives for the European Project", Working Papers in Technology Governance and Economic Dynamics, 59.

Saltelli, A. and Funtowicz, S., 2014, "When all models are wrong: More stringent quality criteria are needed for models used at the science-policy interface", *Issues in Science and Technology*, Winter: 79-85.

Saltelli, A., Guimarães Pereira, Â., van der Sluijs, J.P. and Funtowicz, S., 2013, "What do I make of your Latinorum? Sensitivity auditing of mathematical modelling", *International Journal of Foresight and Innovation Policy*, 9(2-4): 213-234.

Sanderson, K., 2013. "Bloggers put chemical reactions through the replication mill", *Nature*, 21 January. doi:10.1038/nature.2013.12262

Sarewitz, D., 2000. "Science and Environmental Policy: An Excess of Objectivity", in Frodeman, R., 2000. *Earth Matters: The Earth Sciences, Philosophy, and the Claims of Community*: 79-98. Prentice Hall.

Sarewitz, D., 2004. "How science makes environmental controversies worse", *Environmental Science & Policy*, 7: 385–403.

Sarewitz, D., 2011. "The voice of science: let's agree to disagree", *Nature*, 478: 7.

Shapin, S. and Schaffer, S., 2011(1985). *Leviathan and the Air-Pump: Hobbes, Boyle, and the Experimental Life*. Princeton: Princeton University Press.

Smith-Spangler C., Brandeau M.L., Hunter G.E., Bavinger J.C., Pearson M., Eschbach, P. J., Sundaram, V., Liu, H., Schirmer, P., Stave, C., Olkin, I., Bravata D.M., 2012. "Are organic foods safer or healthier than conventional alternatives? A systematic review", *Annals of Internal Medicine*, 157: 348–366.

Springer, 2015. "Springer and Université Joseph Fourier release SciDetect to discover fake scientific papers". https://www.springer.com/gp/about-springer/media/press-releases/corporate/springer-and-universit%C3%A9-joseph-fourier-release-scidetect-to-discover-fake-scientific-papers--/54166

Star, S. and Griesemer, J., 1989. "Institutional Ecology, 'Translations' and Boundary Objects: Amateurs and Professionals in Berkeley's Museum of Vertebrate Zoology, 1907-39", *Social Studies of Science*, 19(3): 387–420.

Steele, R. D., 2014. *The Open- Source Everything Manifesto – transparency, truth and trust*. Berkeley: North Atlantic Books.

Strumsky, D., Lobo, J. and Tainter, J. A., 2010. "Complexity and the productivity of innovation", *Systems Research and Behavioral Science*, 27(5): 496-509.

Taleb, N. N., 2007. *The Black Swan: The Impact of the Highly Improbable*. New York: Random House.

Taleb, N. N., 2012. *Antifragile: Things That Gain from Disorder*. New York: Random House.

Tallacchini, M., 2009. "From Commitment to Committees", *Seminar*, 497, May: 41-45.

Tallacchini, M., 2015. "To Bind or Not To Bind? Ethics in the evolving EU regulation of emerging technologies" in Hilgartner, S., Miller, C. and Hagendijk, R. (eds.), *Science and Democracy: making knowledge and making power in the biosciences and beyond*. New York: Routledge.

U.S. Holocaust Memorial Museum. "Nazi Medical Experiments".
http://www.ushmm.org/wlc/en/article.php?ModuleId=10005168

Van Noorden, R., 2011. "Science publishing: The trouble with retractions", *Nature*, 478: 26-28.

Wilsdon, J., 2014. "Evidence-based Union? A new alliance for science advice in Europe", *Guardian*, 23 June.
http://www.theguardian.com/science/political-science/2014/jun/23/evidence-based-union-a-new-alliance-for-science-advice-in-europe

Winner, L., 1986. *The Whale and the Reactor: A Search for Limits in an Age of High Technology*. Chicago: University of Chicago Press.

Wynne, B, 1993. "Public uptake of science: a case for institutional reflexivity", *Public Understanding of Science*, 2(4): 321-337.

5

NUMBERS RUNNING WILD

Jeroen P. van der Sluijs

Introduction

This chapter[1] is about craft skills with numbers and, in particular, about problems with the use of numbers of unknown pedigree. As an example, I will discuss a very striking number that appeared in the mainstream press in early May 2015: a new scientific study was reported to have found that 7.9% of species would become extinct as a result of climate change.[2] What was quite remarkable about this number is that it had two digits: not 10%, not 8%, but precisely 7.9% of species were to suffer for humanity's carbon sins. A question we will address later is

---

[1] This chapter is adapted from a talk given at "Significant Digits: Responsible Use of Quantitative Information", a workshop organized by the Joint Research Centre of the European Commission at the *Fondation Universitaire* in Brussels on 9-10 June 2015. See https://ec.europa.eu/jrc/en/event/conference/use-quantitative-information

[2] For instance, in the *New York Times*: "Overall, he found that 7.9 percent of species were predicted to become extinct from climate change" (Zimmer, 2015).

the kind of distribution we might expect around this figure of 7.9%.

When a number from the domain of science is catapulted into a user community (for instance, a community of policymakers or conservation campaigners) that has no idea where it came from and no idea of its pedigree, a great deal can go wrong. We have a classic example of this from climate change research, with the concept of climate sensitivity (Van der Sluijs *et al.*, 1998).

Climate sensitivity is a metric for what happens to the Earth's average surface temperature if the atmospheric CO_2 concentration doubles compared to pre-industrial times. There were some model calculations in the early 1980s, with one model giving an outcome of a 1.5°C rise in temperature, another giving 4.5°C, and several models giving intermediate outcomes. This range of outcomes was passed over to the impact assessment community (who do calculations on, for example, rises in sea level), who had no idea where the range came from and therefore failed to fully grasp its meaning. They interpreted the range as a confidence interval—specifically, as a 95% interval with a best and a worst case—so the challenge was thus to protect human societies against a rise in sea level that corresponded to a worst-case temperature rise of 4.5°C.

However, that range had nothing to do with a confidence interval; it was simply a range of point outcomes of individual climate models whose reliability was completely unknown. We will come back to this later. However, imagine what would happen if the 7.9% species extinction rate was presented confidently at the next United Nations climate change summit. What should negotiators and policymakers do with such a number?

Perhaps the negotiators would agree to set an 'acceptable' rate of extinction. A typical political compromise might be half of what would be expected in a business-as-usual scenario. The quantitative policy target would thus be set at a maximum 3.95% (half of 7.9%) extinction rate. The level of precision has now moreover increased to an inconceivable three significant digits, which, given the gross absence of numerical craft skills and literacy, is not uncommon at the science–policy interface.

The rest of the present chapter is structured in three parts. The first concerns the background to the enterprise of producing numbers for policy and the problems and challenges involved. Then we will briefly introduce the NUSAP method, which is a systematic way of exploring the unquantifiable parts of uncertainty associated with these types of numbers. The final section of the paper will be about the case of extinction risk from climate change, to go into a little detail about the kind of number it is and where it comes from.

The phenomenon of scientific uncertainty

We commonly find ourselves in real-world situations that can be called 'post-normal'; that is, where decisions are urgent, stakes are high and values are in dispute (Funtowicz and Ravetz, 1993). In such situations we cannot afford to wait—we are compelled to make immediate decisions based on very imperfect information, marked by irreducible and largely unquantifiable uncertainty. The values of the various stakeholders may be in conflict. Usually, the way we produce knowledge in situations like this is quite different to how it is done in the mono-disciplinary natural sciences. There is no reproducible laboratory experiment or measurement in the field, so we typically use simulation models with future scenarios, based on all kinds of assumptions and with very serious limits to validation and even to parameter estimation. There are often many hidden problems in these models that have not yet been systematically exposed or systematically critically reviewed. We thus need more of what we call 'knowledge quality assessment methods'.

Let us discuss two ways of depicting uncertainty in the climate sciences. The first is to imagine a cascade of uncertainties in the causal chain of climate change. It might start with the drivers, such as population growth, *etc.*, and then progress to energy futures (what kinds of energy will be used to meet the energy demands of the future generations), leading to all kinds of fuel mixes, each with different levels of CO_2 emissions. There are also the non-CO_2 greenhouse gases whose sources change over time, which then lead to changes in emissions of these greenhouse gases, while their atmospheric fate depends on complex atmospheric chemistry in which the presence of one greenhouse gas can impact on the atmospheric half-life of another. A part of the CO_2 and non-CO_2 greenhouse gas emissions remains in the atmosphere and the rest is redistributed, so we need some modelling of carbon cycles and atmospheric chemistry, adding another layer of uncertainty to the cascade. That then produces some radiative forcing of the climate system; and there is also climate sensitivity, which is still to a large degree uncertain. We can then do some impact modelling. We also often need regional projections to inform regional and local decisions on adaptation, with all the challenges and uncertainties associated with down-scaling — and so on.

Moving in this way, from population growth to energy futures to carbon emissions to climate change to impacts such as the loss of biodiversity, seems to imply that we know the structure of the models in each step of the cascade — that it is just a matter of inputting the right parameter values in order to do the right calculations and even to quantify the uncertainty in the model-chain outcome for local climate impacts. However, do we really know the structure of this complex system well enough to make a reliable conceptual model to predict its behaviour? Are all the assumptions we make to link the models valid?[3]

A second way of understanding uncertainty in earth system modelling presents itself upon reflection on the implications of the geological record of the atmospheric composition of CO_2 and methane over a period of half-a-million years. From the gas bubbles trapped in ice in the Vostok ice core drillings, scientists have been able to reconstruct the composition of the atmosphere in the past and have found that it varied from between 180 to 280 parts per million (ppm) by volume for CO_2. The projections by the International Panel on Climate Change (IPCC) are that this level could rise to 1,100 ppm or so, while we already live in a world in which 400 ppm of CO_2 have been exceeded. That is far outside the range of our present knowledge of this complex system, which means that *the past is no longer the key to the future*. All kinds of new feedback processes may be at work in this system, which has been stressed far beyond its long-term equilibrium. We are basically sailing into *terra incognita*. The late Roger Revelle (Revelle and Suess, 1957) called this "man's greatest geophysical experiment". We simply do not know how the planet will respond under these pressures, yet we continue to produce crisp numbers such as 7.9%.

How does the science–policy interface cope with uncertainties? Two strategies seem dominant: concerned scientists over-sell certainty to support political action, while other groups in society over-emphasise uncertainty to delay or prevent it. Scientists are afraid that if they are too open about uncertainties, there will be a policy stalemate, and they will furthermore be reprimanded for not having done their homework correctly. Policymakers generally

---

[3] On this point, see, for example, Van der Sluijs and Wardekker (2015) and Saltelli *et al.* (2015).

expect that, with enough hard work, science can and will produce the ultimate answer.

This is the common *paralysis by analysis* pitfall. To avoid it, scientists have started to be less open about uncertainty and to exaggerate certainty and consensus in order to advance political decision making. Both overselling certainty and exaggerating uncertainty are strategies driven by a compulsion to influence the policy process. The problem is that neither strategy is adequate to the challenges posed by the nature of the uncertainties we face. As research progresses, it produces surprises: the more we know, the more we are aware of complexities in the system, which may completely change what we expect in the future. We may therefore need far more flexible policies than we might imagine on the basis of certainty.

Another famous, almost iconic, illustration of the problem of uncertainty is found in a real case concerning the protection of a strategic groundwater resource in Denmark, near Copenhagen. The five best scientists in the country were given the same question: to determine the most vulnerable part of the area, *i.e.* the part which would need the highest level of protection against nitrate pollution from intensive agriculture and land use practices. All the scientists used the same basic data and sets of measurements from the area, but they each used different models and approaches for their assessment.

Figure 1. Model predictions on aquifer vulnerability towards nitrate pollution for a 175 km² area west of Copenhagen

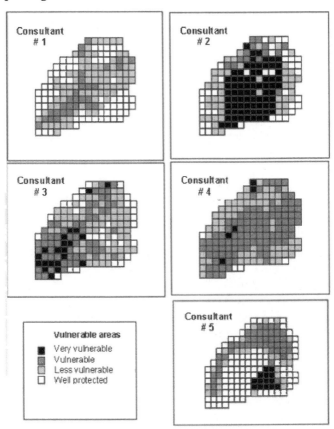

Source: Copenhagen County (2000); Refsgaard *et al.* (2006).

The five vulnerability maps in Figure 1 show the results. Unfortunately, they are not in agreement; they are, in fact, contradictory. However, imagine that science has spoken and that it is now up to policymakers to make a wise decision. What can they do? They can say, "Let us be precautionary and assume the worst case, since if we guard against the worst case, then we are more or less sure that we can protect the zone". But the cost of that strategy may be disproportionately high. They may say, "We need more information before we can decide". That is the infamous *paralysis by analysis* pitfall: more information produces more contradictions, so the decisions are postponed in an infinite loop of evidence-gathering and indeterminate analysis.

So the best science can produce a plurality of scientific perspectives, based on various scientific models and styles of reasoning. It seems that we need to understand better what uncertainty is and why outcomes do not converge. If we look carefully at what is going on in the science–policy interface, we see that the phenomenon of uncertainty is understood in three different ways (Van der Sluijs, 2012). The first is the deficit view, in which uncertainty is seen as a temporary problem. For the time being the science is imperfect, and all we need to do is to collect better data, improve our models, and then, ultimately, science will provide certainty. That is, we reduce uncertainties until we have the precise answer. If that does not work, we just quantify the remaining uncertainty by some confidence interval, error bar, or similar. We use statistical tools such as Monte Carlo simulation, Bayesian belief networks, *etc*. But there is a problem with these tools: we need to feed them with numbers. We need to specify uncertainty ranges for all the model parameters. And if we do not know the range or the distribution and there are no data, we just take a knife, put it to the throat of an expert and ask for a number; we call this 'expert elicitation'. When pushed hard enough, experts will produce a number range or a distribution based on the best of their knowledge, even if such knowledge is illusory. We then put this number range or distribution into the Monte Carlo tool, do the necessary calculations and produce a quantified outcome which fits nicely in a spreadsheet. This corresponds to the 'speaking truth to power' model of interfacing science and decision making. It assumes that we need to produce a quantitative answer, because that is what we believe science is able and supposed to do. And where there is uncertainty, we just speak truth-with-error-bars to policy.

The second understanding of uncertainty is the evidence evaluation view. This is a more pragmatic approach to the problem that science speaks with multiple voices. We see this, for instance, in the way the IPCC works. Experts from different disciplines are brought together to work on a consensus, which means finding the highest common denominator in the science, the most robust claims that we can support with all the published and peer-reviewed studies to date. That is, we know we cannot achieve truth, so we substitute it with a proxy for truth, which is scientific consensus. The model of interfacing science and decision making here is 'speaking consensus to power and to policy'.

These two views can work and be successful in some problem domains, but there are classes of problem where they manifest serious limitations. The first view has the limitation that not everything can be quantified and that experts are forced to quantify uncertainty through best-guessing, because there is no other way. There are many other deficiencies due to model structure uncertainty, the simplification of complex reality in simulation models, the impact of the choice of system boundaries on the outcome, which can never be captured in numeric ranges, and even the assumptions modellers inescapably have to make (*e.g.* Van der Sluijs *et al.*, 2008). The second view has the problem that the consensus approach ignores the unknown probability of high impact scenarios in the risk assessment and tends to confuse unknown probability with negligible probability (*e.g.* Van der Sluijs, 2012). In the first IPCC report, for example, the policymakers' summary mentioned neither the possible collapse of the Antarctic ice sheet nor the collapse of thermohaline circulation in the oceans as relevant scenarios for climate policy. This was not because there was no knowledge; indeed, several published studies had identified those non-linear risks. The detailed chapters of the first IPCC report did review this literature, but it was absolutely impossible to achieve any robust conclusions on these scenarios, so they were excluded from the consensus and did not make it into the policymakers' summary. However, such scenarios were and still are extremely policy-relevant, because policymakers should be interested in the possibility that a five-metre sea level rise could result if the West Antarctic ice sheet were to collapse. Policymakers and society need to know about these types of scenarios even if scientists cannot reach consensus on them.

We therefore need another model, which is the post-normal or complex system approach to uncertainty, in which we acknowledge that uncertainty is permanent and intrinsic to complex, open-ended systems, and that uncertainty can in fact be the result of the way we produce knowledge. For instance, when we construct and use computer simulation models, we cannot avoid making assumptions. There is no way of getting reality into computer code other than by making assumptions, the validity of which can never be thoroughly checked. We thus need another means of controlling the quality of assumptions in models. That requires a more open way of dealing with the deeper dimensions of uncertainty, *etc.*, and is why we have developed tools for knowledge quality assessment (Van der Sluijs *et al.*, 2008, Saltelli *et al.*, 2013).

From this post-normal perspective, the whole science-policy interface changes. It is no longer speaking truth or consensus to power; it is now working deliberatively within imperfections. We cannot always produce the ultimate answer; rather, we have to work within imperfections and we have to do that in dialogue between science, policy and society.

By studying the science–policy interface in different fields — *e.g.* air quality, climate change, biodiversity — we see how differently these types of problems may be solved. A Bayesian approach is often used, where the five different results in the Danish groundwater resource case described above are seen as informative priors that are simply averaged and the resulting model updated if new information becomes available. But what if there is no data, and decisions are needed urgently? What is the quality of the priors? We could take the IPCC approach, in which scientists are locked in a room and not released until they reach a consensus. We could take a precautionary approach and assume the worst case. We could take an academic/bureaucratic approach, whereby the scientist with the highest Hirsch index is given the greatest credibility. We could give preference to the scientist we trust most or the one most in line with our policy agenda, which is what often happens. We could forget the science and decide on an entirely different basis. Alternatively, we could explore the relevance of our ignorance in a post-normal attitude, whereby we try collectively to find wiser ways to deal with uncertainties and avoid the pitfall of taking decisions under the illusion of having tamed uncertainty.

The first view of uncertainty as a temporary state has been very persistent. In a quotation from the first IPCC report in 1990 (IPCC, 1990; see Box 1a), the authors write that there are many uncertainties in their predictions, particularly with regard to the timing, magnitude and regional patterns of climate change due to incomplete understanding of sources and sinks of greenhouse gases, clouds, oceans and polar ice. They go on to say that these processes are already partly understood, and that they are confident that the uncertainties can be reduced by further research. That was a strong claim in 1990 for the scientific community working on climate change; it is what they believed.

Box 1.

Box 1a. IPCC 1990 optimism about reducing uncertainty.	Box 1b. Former IPCC chairman, the late Bert Bolin, on the objective to reduce climate uncertainties.
"There are many uncertainties in our predictions particularly with regard to the timing, magnitude and regional patterns of climate change, due to our incomplete understanding of: • sources and sinks of greenhouse gases, which affect predictions of future concentrations • clouds, which strongly influence the magnitude of climate change • oceans, which influence the timing and patterns of climate change • polar ice sheets which affect predictions of sea level rise These processes are already partially understood, and we are confident that the uncertainties can be reduced by further research However, the complexity of the system means that we cannot rule out surprises." (IPCC, 1990: xii)	"We cannot be certain that this can be achieved easily and we do know it will take time. Since a fundamentally chaotic climate system is predictable only to a certain degree, our research achievements will always remain uncertain. Exploring the significance and characteristics of this uncertainty is a fundamental challenge to the scientific community." (Bolin, 1994)

Just before the second IPCC assessment report (1995) was issued, the late Bert Bolin, then chairman of the IPCC, said in a conference that they could not be certain that uncertainty could be easily reduced. He continued by saying it would take time to reduce uncertainty, in line with the deficit view of temporary nature of uncertainty. But he concluded that, since the fundamentally chaotic climate system is predictable only to a certain degree, there were limits to predictability, and research results would always remain uncertain. He thus finished by adopting the third view we have identified, namely that uncertainty is permanent and will never go away.

Bolin added that exploring the significance and characteristics of this uncertainty was a fundamental challenge to the scientific community. Most people would say that the fundamental challenge to science is to determine facts and achieve certainty, so in this respect Bolin was being 'post-normal', even in the early stages of post-normal science.

Let us look more closely at the IPCC's ambition in 1990 to reduce uncertainty in various domains (Box 1a). The first item they listed was "sources and sinks of greenhouse gases". Ten years later, what we know about sources and sinks of CO_2, only one of various greenhouse gases, is illustrated in Figure 2.

Figure 2. Emission scenarios produced by six different energy models (Maria, Message, Aim, Minicam, ASF, Image), each using four harmonized sets of scenario assumptions, as presented in the IPCC Special Report on Emissions Scenarios

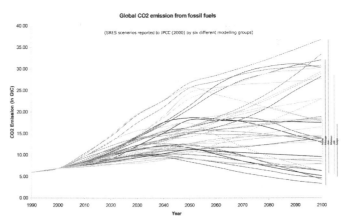

Source: IPCC (2000).

Four different sets of assumptions, A1, A2, B1 and B2, were used in each of six models presented in the IPCC Special Report on Emissions Scenarios (2000), corresponding to a focus on economy (A) versus environment (B) and on development and globalization (1) versus regionalization (2) of the world. These scenario families ensured harmonized assumptions on population growth, economic growth, and technology transfer. Despite this, there were very wide discrepancies in what the six models projected for these same four sets of scenario assumptions. Apparently not much has been reduced in terms of uncertainty about sources and sinks of CO_2 in ten years of research.

Now let us look at the key parameter of climate change, so-called climate sensitivity, 25 years after the first IPCC report predicted a reduction in uncertainties. As indicated above, climate sensitivity is a measure of the average global warming that would be produced by a doubling of the CO_2 concentration (to 560 ppm) compared to the pre-industrial level of 280 ppm. The first synthesis estimate of climate sensitivity was made in 1979, when the U.S. National Academy of Sciences made an assessment that combined results from several general circulation models. The outcomes ranged from 2 to 3.5°C; with some expert wisdom the range was widened, in acknowledgement that the models were imperfect. By including this tacit knowledge of uncertainties in the models, they arrived at the conclusion that the anticipated average rise in temperature must be somewhere between 1.5 and 4.5°C, and their best guess was 3°C.

Table 1. Evolution of knowledge on climate sensitivity over the past 35 years

Assessment report	Range of GCM results (°C)	Concluded range (°C)	Concluded best guess (°C)
NAS 1979	2-3.5	1.5-4.5	3
NAS 1983	2-3.5	1.5-4.5	3
Villach 1985	1.5-5.5	1.5-4.5	3
IPCC AR1 1990	1.9-5.2	1.5-4.5	2.5
IPCC AR2 1995	MME	1.5-4.5	2.5
IPCC AR3 2001	MME	1.5-4.5	Not given
IPCC AR4 2007	MME	2.5-4.5	3
IPCC AR5 2013	MME (0.5-9)	1.5-4.5*	Not given

Source: Updated from Van der Sluijs *et al.* (1998).

* "Likely" (17-83%) range. Note that prior to Assessment Report 4 (AR4), ranges were not clearly defined. MME = Multi Modal Ensemble.

A couple of years later the National Academy of Sciences (NAS) updated the assessment with new outcomes from models; thereafter came the Villach conference in 1985, where the published literature showed models with outcomes as low as 1.5°C and as high as 5.5°C (see Table 1). However, something strange then occurred, because the range given in the conclusions of the Villach conference was narrower than this. The NAS had widened the range in their assessment because the experts took into consideration the uncertainties in the model. But in the new assessment in Villach in 1985, the range resulting from the inventory was narrowed by the experts, because they did not want to grant too much credibility to one (perceived) outlier model. In effect, the argumentative chain linking the set of published literature to the recommended range completely absorbed all the changes in the science and all the uncertainties, to keep the scientific basis for policy making stable.

From the second IPCC report onwards, multi-model perturbed physics ensemble modelling was used, in which each model produced a range or distribution which was then combined into an ensemble. The fifth report (IPCC, 2013) suddenly has a footnote on the recommended range, so the multi-model ensemble produces numbers between half a degree and 9°C for climate sensitivity; the recommended range is still 1.5 to 4.5°C; and the footnote indicates that this is to be interpreted as the likely range, the 17th to 83rd percentile of the distribution.

The modellers involved in the first assessment report said in an interview that the range recommended in that report was to be seen as the equivalent of a 95% interval (Van der Sluijs *et al.*, 1998), whereas in the fifth report that same 1.5-4.5°C range has suddenly become a 66% interval. In other words, the meaning of the range has changed in order to keep it constant over time for use in the science-policy interface.

Looking at the distributions from different sources (models, paleo records, *etc.*) (Figure 3), we see some ranges, such as the likely 1.5 to 4.5°C range from Chapter 12 of the fifth assessment report of the IPCC. More than a century ago, Svante Arrhenius (1896) did the first calculations on the potential doubling of CO_2 and came up with 4.95°C at the Equator, rising to 5.95°C at 60°S and to 6.05°C at 70°N. The IPCC's fifth assessment says the temperature rise is very unlikely to be greater than 6°C, so even the one-hundred-year-old estimate without computer models, with crowd sourcing to students for the calculations, is still well in the range of what IPCC now considers possible. Have uncertainties diminished, as the IPCC foresaw in 1990? On the contrary, they seem to have increased, at least in this key climate parameter.

Figure 3. Probability density functions, distributions and ranges for equilibrium climate sensitivity

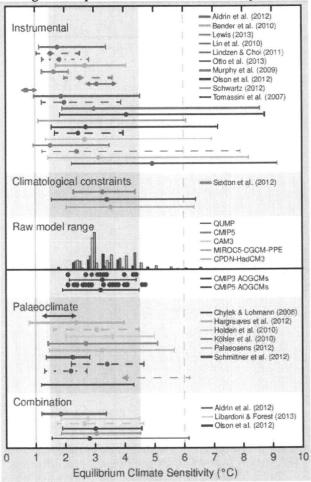

Grey shaded range: likely 1.5°C to 4.5°C range. Grey solid line: extremely unlikely, less than 1°C. Grey dashed line: very unlikely, greater than 6°C.

Source: Reprinted from Box 12.2, Fig. 1, from Collins *et al.* (2013).

In addition to asking the models, one can also ask the experts, as Morgan and Keith (1995) did. Looking at subjective assessments of climate sensitivity (Figure 4) by the top 16 climate modellers (based on their publication record in the field) in the USA, we see that some cannot even exclude the possibility that it is a negative number—that is, that the equilibrium temperature change will be smaller than zero due to some feedback that overcompensates for the warming by cooling in the long run. Overall, we see that there are many ways of assessing climate sensitivity that lead to very different views.

Figure 4. Box plots of elicited probability distributions of climate sensitivity, the changes in globally averaged surface temperature for a 2 X [CO₂] forcing

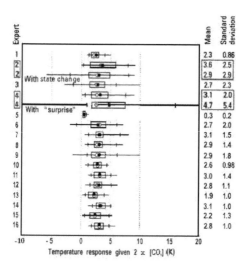

Horizontal line denotes range from minimum to maximum assessed possible values. Vertical tick marks indicate locations of lower 5th and upper 95th percentiles. Box indicates interval spanned by 50% confidence interval. Solid dot is the mean and open dot is the median. The two columns of numbers on the right side of the figure report values of mean and standard deviation of the distributions.

Source: Morgan and Keith (1995).

The case of species extinction from climate change

We return now to the number 7.9% with which this paper began, namely the predicted rate of extinction of species due to climate change. Models have produced widely varying estimates of the anticipated extinction of species. Urban (2015) assembled 131 studies. At first sight, 131 is an easily understandable number. It is just a count of the number of studies. But even that number is problematic, because some studies combined a number of other studies, but were counted as one, while there was also some overlap due to the fact that any given study may have combined several studies that were considered separately elsewhere, and so on.

Urban's meta-analysis, published in *Science*, found that, overall, these 131 studies predicted that 7.9% of species would become extinct due to climate change. He also provided a confidence interval: for 2°C it would be 5.2% and for 4.3°C it would be 16% of all species. Linguistically Urban is literally talking about a percentage of "all species". The opening paragraph of the study that presented this 7.9% tells us that the goal of the whole exercise was to inform international policy decisions about the biological cost of failing to curb climate change and to support specific conservation strategies to protect the most threatened species.

In practice, here is what Urban did. He considered 131 studies and classified them according to their outcome in terms of predicted risk of extinction, with one category of zero (containing 30 studies), another for zero to 5% (containing another 30 studies), then 5% to 10% (containing about 14 studies), *etc*. Then we see an average, 7.9%, summarizing all of these studies. It is not clear what year this number refers to. Each of the 131 studies had a different reference period, so no year is mentioned in the paper. Zero to 5% includes zero, so one should stack the two bars for the first category in Figure 5 (60 studies), making for a very asymmetrical distribution. The figure 7.9% does not give any useful information about this distribution at all. It is misleading to communicate the average of such a skewed distribution without also giving the distribution; in addition, we should know where all these numbers come from.

Figure 5. Histogram of percent extinction risk from climate change for 131 studies

Overall extinction risk = 7.9% (95% CI: 6.2, 9.8)

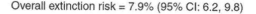

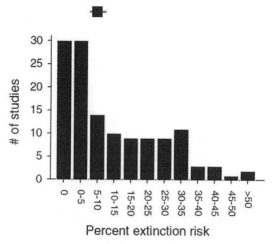

Source: Urban (2015).

Systematic critical reflection on uncertainty

Before moving on to another estimate of climate extinction risk from Thomas *et al.* (2004) in *Nature*, let us look at how we should critically reflect on uncertainties in these types of studies. This is the guidance approach (Figure 6) that the Netherlands Environmental Assessment Agency developed jointly with a trans-disciplinary group of scientists in the field of uncertainty, with backgrounds ranging from the policy sciences and the humanities to the natural sciences. It starts from the idea that we need critical reflection on all the phases of knowledge production, from the framing of the problem and the drawing of the system boundaries in making our models, to what is included and excluded—whether, for example, we should treat climate change separately from ocean pollution, air pollution, *etc.*, or rather try to include all these complex interactions.

Figure 6. The Dutch Guidance approach to systematic reflection on uncertainty and quality in science for policy

Foci and key issues in knowledge quality assessment (ref. 9)

Foci	Key issues
Problem framing	Other problem views; interwovenness with other problems; system boundaries; role of results in policy process; relation to previous assessments
Involvement of stakeholders	Identifying stakeholders; their views and roles; controversies; mode of involvement
Selection of indicators	Adequate backing for selection; alternative indicators; support for selection in science, society, and politics
Appraisal of knowledge base	Quality required; bottlenecks in available knowledge and methods; impact of bottlenecks on quality of results
Mapping and assessing relevant uncertainties	Identification and prioritisation of key uncertainties; choice of methods to assess these; assessing robustness of conclusions
Reporting uncertainty information	Context of reporting; robustness and clarity of main messages; policy implications of uncertainty; balanced and consistent representation in progressive disclosure of uncertainty information; traceability and adequate backing

Source: Van der Sluijs *et al.* (2008).

We always simplify in scientific assessments. We tend to set very limited system boundaries in order to keep scientific assessments manageable, but we have to understand the impact of these design choices on the validity and scope of the conclusions of such assessments. Stakeholder engagement is ever more important. Stakeholders are crucial in co-framing the problem and co-deciding what is relevant to address; they can provide useful information and data that scientists have otherwise no access to; and they can be a critical resource in quality control and extended peer review.

Next, there is the selection of indicators. In scientific assessments all kinds of indicators are used: for instance, the percentage of species that are at risk of extinction or the anticipated temperature rise due to the doubling of CO_2 concentrations. Are these indicators relevant to the policy challenges that we face? Are there alternatives? Are they chosen because a particular model is available to use, or were they designed to address a particular societal problem? Usually, scientists attempt to answer the questions of policymakers by using an existing model or toolkit that is on the shelf but which does not really match the decision-making needs.

Then there is the appraisal of the knowledge base, meaning that we have to systematically look at all the problems and limitations in the available knowledge and characterize it in terms of its pedigree, strengths and weaknesses. We will return to the concept of pedigree later.

Next, we have to map all the sources and relevant types of uncertainty, to be aware of where they are, what they are, and how to take them into consideration in the policy advice; we then have to report the information on uncertainty in a way appropriate to the decision-making context, so that decision-makers can actually use it to make more robust, more resilient or more flexible decisions that take these uncertainties into account in a more sophisticated way than is currently done.

The core message is that uncertainty is much more than a number and that it has dimensions. Three of these are defined in the book *Uncertainty and Quality in Science for Policy,* the 1990 classic by Funtowicz and Ravetz. There are technical, methodological and epistemological dimensions of uncertainty in numbers. The technical dimension corresponds to inexactness, error bars, *etc.*, while the methodological dimension corresponds to unreliability (for instance, whether the model structure is reliable). The epistemological dimension corresponds to the borders with ignorance and the limits of our ability to know and understand complex systems. We could argue that there is another dimension, namely the societal robustness of the knowledge, but this could also be considered to belong to the epistemological category.

Figure 7. Successive recommended values of the fine structure constant α^{-1}

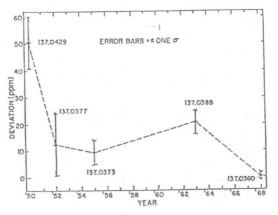

Fig. 1.   Successive recommended values of the fine-structure constand α^{-1} (B. N. Taylor *et al.*, 1969, 7)

Source: Taylor *et al.* (1969).

An example from Funtowicz and Ravetz (1990) concerns the recommended number for a physical constant, the *fine structure constant*. Recommended values reported in successive editions of the *Handbook of Chemistry and Physics* — the bible of the natural sciences — have changed over time. This is because science progresses. There are three items in the notational system used in the *Handbook*:

a number, a unit and a spread, the latter usually reported as a standard deviation. If we look at the number and the standard deviations that are recommended in successive editions of the *Handbook* for the fine structure constant, we see something strange.

If there were only measurement error with random distribution to be considered, we would expect that 95% of the distribution would be captured by two standard deviations around the mean, and almost everything captured by plus or minus three standard deviations. However, we see that in 1968 the recommended value and reported error bar (one standard deviation) are far outside the error bar reported in 1950; the recommended values in 1950 are more than four standard deviations apart. So apparently there is more to say about uncertainty than can be captured in this standard deviation.

This was perhaps the first time the fine structure constant had been measured, but we know that systematic error means that no two laboratories will produce exactly the same result. If users of this number could have been warned early on that it had some limitations and that the reported standard deviation could not capture all the uncertainty, it might have helped to avoid misunderstandings of this type of number. This kind of uncertainty has somehow to be communicated.

This example was of a physical constant, which may not capture the imagination of the public. Let us consider a policy relevant number, the emission levels of ammonia, an air pollutant mainly coming from the agricultural system and for which emission reduction policies are in place in the Netherlands. Intensive cattle breeding in the Netherlands produces a great deal of NH_3 (ammonia) air pollution. Emission reduction policies are often relative to a reference year. Let us take 1995 as the reference year, and imagine that we want to reduce emissions by, for instance, 10% relative to that year. We take the Dutch "State of the

Environment Report"[4] (RIVM, 1996, 1997, 1998, 1999, 2000, 2001, 2002), which has tables for all the environmental indicators for which the Netherlands has implemented policies. We look up the 1995 emissions levels in the 1996 edition of the report, and it is slightly more than 150 million kilograms. Then we look up the same number in more recent editions of the report. In the 1997 edition there were some recalculations and there is a new number for the 1995 emission, which is slightly lower (see Figure 8). Nothing changes in the 1998 edition, but there is a jump in the 1999 edition: it seems that knowledge of the emissions in 1995 has suddenly changed, and we see now that the 1995 level is almost 30% higher than we thought a couple of years earlier. The 30% change in the number due to the recalculation is three times larger than the hypothetical long-term policy objective of 10% reduction in emissions, which is discomfiting because it has serious policy implications.

What happened here? We might imagine that someone goes out to measure gas emissions every year, but that is not the case. The calculation of NH_3 emissions in a given year is the product of an agricultural model that follows the nitrogen, starting with the animal feed and taking into account the number of cows, chickens and pigs in the Netherlands. These animals are kept in different agricultural subsystems, some in barns, some in meadows. The model assumes partition coefficients of how much nitrogen from the food ends up in the manure. A fraction of the nitrogen in the manure in a barn or in a meadow is emitted as NH_3 into the air. This is then calculated for each type of cow, chicken and pig in each agricultural subsystem and finally everything is summed to serve as an

---

[4] "Milieubalans", an annual publication by the Netherlands Environmental Assessment Agency, http://www.pbl.nl, which was formerly part of RIVM (National Institute for Public Health and the Environment).

estimate of the total NH_3 emissions in a given year. So the numbers of animals in 1995 considered in the calculation remained constant, and what changed in the recalculation in the 1999 edition was the assumed emission factor of the biggest type of barns used in the Netherlands. A small change in the emission factor of the largest pig farms makes a big difference in the outcome for the model-calculated emissions.

Figure 8. Total NH_3 emissions in the Netherlands in 1995 as reported in successive editions of the "State of the Environment Report" of the Netherlands Environmental Assessment Agency

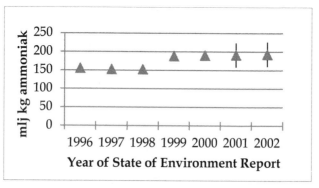

Uncertainty as discussed in the present chapter may appear to be the preoccupation of a small community of practitioners. In fact, its relevance to policy making may sometimes cause it to hit the headlines. In 1999 there was a scandal at the Netherlands Environmental Assessment Agency when whistle-blower Hans de Kwaadsteniet revealed to the media the use of poorly validated models for the production of environmental assessments (Van der Sluijs, 2002). As a consequence of the scandal and its follow-up, it became mandatory for the Environmental Assessment Agency to report uncertainties. From the 2001 edition of the "State of the Environment Report" onwards error bars are reported, reflecting two standard deviations. What is striking is that the size of these error bars is smaller than the change observed as a result of the recalculation in the 1999 edition. In terms of the different types of uncertainty, that means that the methodological uncertainty (recalculation with updated assumption) was more significant here than the technical uncertainty (error bar).

Qualified quantities: the NUSAP approach

What about the third dimension, the borders with ignorance, which might be even more important? How do we assess that dimension of uncertainty? We need more qualifiers of scientific information, and that is where the new notational system comes in. The classic notational system only has room for a number, a unit and a spread, whereas for all the problems in the post-normal domain, we may need to issue something like a patient information leaflet that warns users about the risks involved in using numbers and provides guidance on how to use them responsibly. To that end, Funtowicz and Ravetz (1990) proposed adding *assessment* and *pedigree* to the notational system, producing the NUSAP system: numeral, unit, spread, assessment, pedigree. Assessment is an expert judgement on reliability that puts the number into context, and pedigree is a systematic multi-criteria evaluation to characterize the status of the scientific information in terms of how it was produced, where it comes from, and how we should understand the pedigree of a given number. This system is sensitive to our proximity to ignorance. Pedigree is assessed with the help of a pedigree matrix, describing pedigree criteria along with an ordinal 5-point scoring scale for each criterion. Table 2 presents how a pedigree may be articulated.

For example, a model may have different parameters, some based on well-established science and others more speculative, and we have to be able to make the distinction between them. In terms of criteria we can use, we have the empirical basis for the number, the theoretical basis behind the parameter, the methodological rigour with which the number was produced, the degree of validation or the degree to which it was a proxy as compared to the thing we really want to know.

Table 2. Example pedigree matrix for emission monitoring data

Score	Proxy	Empirical basis	Methodological rigour	Validation
4	An exact measure of the desired quantity	Controlled experiments and large sample, direct measurements	Best available practice in well-established discipline	Compared with independent measurements of the same variable over long duration
3	Good fit or measure	Historical/field data uncontrolled experiments, small sample, direct measurements	Reliable method common within est. discipline, best available practice in immature discipline	Compared with independent measurements of closely related variable over shorter period
2	Well correlated but not measuring the same thing	Modelled/derived data, indirect measurements	Acceptable method but limited consensus on reliability	Measurements not independent, proxy variable, limited period
1	Weak correlation but commonalities in measure	Educated guesses, indirect, approx. rule of thumb estimate	Preliminary methods, unknown reliability	Weak and very indirect validation
0	Not correlated and not clearly related	Crude speculation	No discernible rigour	No validation performed

Source: Van der Sluijs *et al.* (2005); adapted from Ellis *et al.* (2000).

For example, the empirical basis can be anything between crude speculation, an educated guess, model-derived data, small samples of direct measurements or, in the best case, a large sample of direct measurements of the thing we seek to know. The level of methodological rigour can range from a state of no discernible rigour to a preliminary method with unknown reliability, all the way to the best available practice in a well-established discipline. Validation can go from weak or indirect validation all the way to comparison with independent measurements. If we have this type of meta-information about the numbers, we can understand that they are not as strong as they might look at first sight and are much more prepared for them to change in the future.

Figure 9. Example of presentation of pedigree scores of monitoring data of air pollutants.

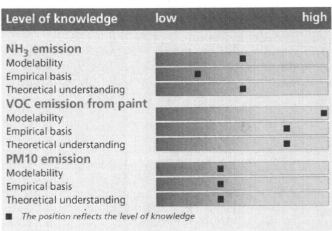

Source: Van der Sluijs *et al.* (2008). © IOP Publishing Ltd. CC BY-NC-SA, doi: 10.1088/1748-9326/3/2/024008

To illustrate, three air pollutants are monitored in a country, but the status of knowledge on how we can best estimate the yearly emissions varies. We have much more knowledge about the volatile organic components of paint, which we can easily model and for which the empirical basis and theoretical understanding are quite strong, than about how particles are formed in a combustion process or about the empirical basis for monitoring ammonia emissions from cattle breeding. This type of information gives a better idea of what types of policies can be formed on the basis of these numbers. Figure 9 gives an example of how this can be communicated.

A paper on extinction risks due to climate change from Thomas *et al.* (2004) serves as another example. It is a four-page paper in *Nature* presenting a very simple model. The study's authors predict, on the basis of a mid-range climate warming scenario for 2015, that 15% to 37% of species in the sample of regions and taxa will be "committed to extinction". This was one of the studies included in the meta-analysis that arrived at the 7.9% figure. The 7.9% in Urban (2015) referred to "species", while Thomas *et al.* refer to "species in their sample of regions and taxa". The meta-analysis talks about species that will become extinct, while this paper talks about species being "committed to extinction", which is a rather vague, poorly defined concept.

There is an interesting claim in the last line of the introductory paragraph of the Thomas *et al.* paper, that these estimates "show the importance of rapid implementation of technologies to decrease greenhouse gas emissions and strategies for carbon sequestration". The logic is not clear. Does the fact that species will go extinct mean that we should design strategies to store CO_2 underground? How do the authors jump from species extinction to a particular preferred solution (carbon sequestration)? This sounds like an opinion for which the underlying arguments are not even given. However, what we are interested in is the quantified outcome of this study, the rate of 15% to 37% of species being committed to extinction under mid-range climate warming scenarios. The authors arrived at this range by applying the so-called species-area relationship from ecology, which says that the number of species depends on the area of the habitat of the species, and that by consequence, if the habitat shrinks, the likelihood of a species becoming extinct increases. The greater the shrinkage of the area, the more species will become extinct. There is a simple relationship between the number of species and the area of a given habitat; its parameters have been estimated based on empirical data:

$$S = c\,A^{\,z}$$

where S = number of species, A = area, c = constant and z ≈ 0.25. We see that this formula yields a number, and if we compare it to climate projections, and the area before and after climate change for each habitat, then we get a ratio which is simply the ratio of the area to the power z, which is itself an empirically derived parameter estimated to be 0.25. There was, however, a sensitivity analysis for the value of z.

The authors collected climate projections for habitat changes from the literature and grouped these into three classes of warming (low, mid-range, high) and explored the outcomes for two extreme assumptions on dispersal (no dispersal and full dispersal). To briefly explain dispersal: as a rule of thumb, for each degree of warming we get roughly a 100-kilometre shift of climate zones from the Equator to the poles, a 150-metre upward shift in the mountains. The species have to catch up with that and migrate (disperse) to the new area with favourable climatic conditions for their survival. Some can migrate more easily than others. A forest needs more time to migrate, because trees cannot walk, so it depends on the speed of the seed cycle. From the paleo-record we know that an oak forest can keep up with 0.12°C per decade, and if the increase in temperature is faster, the oak forest will die back more rapidly on one side than it will expand on the other to populate the new zone where the climate is now favourable.

Therefore, the climate zones and associated habitats shift poleward with climate change, but if there is no dispersal, species can only survive in the overlapping area between the old and the new favourable climate zones, while if there is full dispersal, the species can migrate completely to the new, favourable area.

We see that the outcome that made it to the conclusion is the mid-range scenario, with the lowest and the highest number in the reported range corresponding to the full and no dispersal assumptions; the resulting range being 15% to 37%. Next, Thomas *et al.* (2004) present a huge table in which they list in aggregated (per taxa and region) form the data from the 1,103 species which they studied, including mammals in Mexico, birds in different regions, frogs, reptiles, butterflies, plants, *etc.* There are three numbers for each scenario with and without dispersal, representing three ways of aggregating the data from individual species in each taxa and region to the whole set of species in that taxa and region.

However, there are many missing numbers in the table. The authors explain in the methods section how they interpolated the missing numbers, which they had to do in order to calculate the bottom line, the extinction risk for all the species in the sample. These aggregated numbers, which are used in the conclusion of the paper, thus include a large number of hidden interpolated numbers.

Overall, we see that there are many assumptions hidden behind the numbers, for example in the interpolation algorithm. There is also a bias in the sample of species that are included in the analysis, because they are species on which publications exist in terms of what climate change does to their habitat. This means that the study may be focused on those species most sensitive to climate change:

such results cannot be extrapolated to all the species on the planet. Considering again the Urban (2015) study that concluded that 7.9% of all species will be made extinct — what would that be in absolute terms? Nobody knows the number of species on the planet. Does it refer to eukaryotic species only (estimated to be 8.7 million plus or minus 1.3 million (Mora *et al.*, 2011)) or to all species on Earth? Urban does not specify, so even the *unit* (the second qualifier in the NUSAP) is ambiguous.

All these aspects of uncertainty become evident in the process of applying NUSAP and the pedigree matrix to systematize critical reflection on the strengths and weaknesses of scientific knowledge claims. The pedigree matrix can have criteria such as proxy, quality and quantity of the empirical basis of this model, theoretical understanding, representation of the underlying mechanisms, plausibility and colleague consensus. As an exercise, my M.Sc. and Ph.D. students apply such a pedigree matrix to the Thomas *et al.* (2004) study. The estimate of the extinction risk is often scored somewhere in the middle for proxy. There is a correlation between area of habitat size and extinction risk, but it is not the same thing; we are not modelling extinction, but area loss, which is assumed to be correlated to extinction. The quality and quantity of the model, which are somewhere between an educated guess and modelled and derived data, attain quite a low pedigree score of between 1 and 2. Regarding theoretical understanding, it is an accepted theory of a partial nature with a limited consensus on reliability, but some students think it is a preliminary theory. The other pedigree criterion also gets quite low scores, based on preliminary methods and weak and indirect validation.

Conclusions

Figures such as the 7.9% that we have discussed in detail in this chapter are often a first attempt to quantify a complex phenomenon, in this case the risk of extinction of species due to anthropogenic climate change. We have treated this number as an example of what we perceive to be a dangerous practice—the production and use of crisp figures to give the impression that science produces truth.

The fact that such hyper-precise quantifications may emerge from statistical averaging processes or from computational algorithms is not a justification for their publication. Numbers ought to be used responsibly; in our view the approach of Urban (2015) leaves much room for improvement. Science—especially when deployed at the science-policy interface—should involve what we call 'craft skills' with numbers. Quantification should be a much more nuanced and reflective process (Porter, 1995; for a discussion see Chapter 2, this volume). A peer review process for quantitative evidence would need to systematically include approaches such as NUSAP, sensitivity auditing in the case of mathematical or statistical models, and in general the exercise of good judgment.

The liberty to quantify should be used with discretion, and should go hand-in-hand with the duty to refrain from quantification when appropriate. The practice of throwing magic numbers into the arena for public consumption is in our view one of the symptoms and causes of the crisis in science that is under analysis in this book, representing, as it does, an abdication of the scientist's responsibility to ensure that the craft of science has been applied with all due diligence.

References

Arrhenius, S., 1896. "On the influence of Carbonic Acid in the Air upon the Temperature of the Ground", *The London, Edinburgh and Dublin Philosophical Magazine and Journal of Science*, 41(251): 237-276.

Bolin, B., 1994. "Science and Policy Making". *Ambio* 23(1): 25-29.

Collins, M., Knutti, R., Arblaster, J., Dufresne, J.-L., Fichefet, T., Friedlingstein, P., Gao, X., Gutowski, W.J., Johns, T., Krinner, G., Shongwe, M., Tebaldi, C., Weaver A.J. and Wehner, M., 2013. Long-term Climate Change: Projections, Commitments and Irreversibility. In: Stocker, T.F., Qin, D., Plattner, G.-K., Tignor, M., Allen, S.K., Boschung, J., Nauels, A., Xia, Y., Bex V. and Midgley P.M. (eds.) *Climate Change 2013: The Physical Science Basis. Contribution of Working Group I to the Fifth Assessment Report of the Intergovernmental Panel on Climate Change*. Cambridge, UK: Cambridge University Press.

Copenhagen County, 2000. "Pilot project on establishment of methodology for zonation of groundwater vulnerability. Proceedings from seminar on groundwater zonation", 7 November. County of Copenhagen (in Danish).

Ellis, E. C., Li, R. G., Yang, L. Z. and Cheng, X., 2000. "Long-term Change in Village-Scale Ecosystems in China Using Landscape and Statistical Methods", *Ecological Applications*, 10: 1057-1073.

Funtowicz, S. O. and Ravetz, J. R., 1990. *Uncertainty and quality in science for policy*. Dordrecht: Kluwer.

Funtowicz, S. O. and Ravetz, J. R., 1993. "Science for the Post-Normal Age", *Futures*, 25(7): 735-755.

IPCC, 1990. "Policymakers' summary", in Houghton, J., Jenkins, G. and Ephraums, J. (eds.), *Climate Change, The IPCC Scientific Assessment*: vii-xxxiv. Cambridge, UK: Cambridge University Press.

IPCC, 2000. *Emissions Scenarios*. Cambridge, UK: Cambridge University Press.

IPCC, 2013. "IPCC Fifth Assessment Report: Working Group I Report", in *Climate Change 2013: The Physical Science Basis*. Geneva: IPCC.

Mora, C., Tittensor, D. P., Adl, S., Simpson, A. G. B. and Worm, B., 2011. "How Many Species Are There on Earth and in the Ocean?", *PLoS Biology*, 9(8): e1001127.

Morgan, M. G. and Keith, D. W., 1995. "Subjective Judgments by Climate Experts", *Environmental Science & Technology*, 29(10): 468A-476A.

Porter, T. M., 1995, *Trust in Numbers. The Pursuit of Objectivity in Science and Public Life*. Princeton: Princeton University Press.

Refsgaard, J-C., van der Sluijs, J. P., Brown, J. D., and van der Keur, P., 2006. "A framework for dealing with uncertainty due to model structure error", *Advances in Water Resources*, 29(11): 1586-1597.

Revelle, R. and Suess, H. E., 1957. "Carbon Dioxide Exchange Between Atmosphere and Ocean and the Question of an Increase of Atmospheric CO_2 during the Past Decades", *Tellus* 9(1): 18-27.

RIVM, 1996. *Milieubalans 1995*. Alphen aan den Rijn: Samsom.

RIVM, 1997. *Milieubalans 1996*. Alphen aan den Rijn: Samsom.

RIVM, 1998. *Milieubalans 1997*. Alphen aan den Rijn: Samsom.

RIVM, 1999. *Milieubalans 1998*. Alphen aan den Rijn: Samsom.

RIVM, 2000. *Milieubalans 1999*. Alphen aan den Rijn: Samsom.

RIVM, 2001. *Milieubalans 2000*. Alphen aan den Rijn: Samsom.

RIVM, 2002. *Milieubalans 2001*. Alphen aan den Rijn: Kluwer.

Saltelli, A., Guimarães Pereira, Â., Van der Sluijs, J.P. and Funtowicz, S., 2013, "What do I make of your latinorum? Sensitivity auditing of mathematical modelling", *International Journal of Foresight and Innovation Policy*, (9), 2/3/4, 213–234.

Saltelli, A., Stark, P.B., Becker, W., and Stano, P., 2015, "Climate Models As Economic Guides Scientific Challenge or Quixotic Quest?", *Issues in Science and Technology*, 31(3).

Taylor, B. N., Parker, W. H. and Langenberg, D. N., 1969. *The fundamental constants and quantum electro-dynamics*. New York: Academic Press.

Thomas, C. D., Cameron, A., Green R. E. Bakkenes, M., Beaumont, L. J. Collingham, Y. C., Erasmus, B. F. N., Ferreira de Siqueira, M. Grainger, A., Hannah, L., Hughes, L., Huntley, B., Van Jaarsveld, A. S., Midgley, G. F., Miles, L., Ortega-Huerta, M. A., Peterson, A. T., Phillips, O. L., and Williams, S. E., 2004. "Extinction risk from climate change", *Nature*, 427: 145-148.

Urban, M. C., 2015. "Accelerating extinction risk from climate change", *Science*, 348: 571-573.

Van der Sluijs, J. P., 2002. "A way out of the credibility crisis around model-use in Integrated Environmental Assessment". *Futures*, 34: 133-146.

Van der Sluijs, J. P., 2005. "Uncertainty as a monster in the science policy interface: four coping strategies", *Water science and technology*, 52(6): 87–92.

Van der Sluijs, J. P., 2012. "Uncertainty and dissent in climate risk assessment, a post-normal perspective", *Nature and Culture*, 7(2): 174-195.

Van der Sluijs, J. P., van Eijndhoven, J. C. M., Shackley, S. and Wynne, B., 1998. "Anchoring Devices in Science For Policy: The Case of Consensus Around Climate Sensitivity", *Social Studies of Science*, 28: 291-323.

Van der Sluijs, J. P., Risbey, J. and Ravetz, J. (2005b). "Uncertainty Assessment of VOC emissions from Paint in the Netherlands", *Environmental Monitoring and Assessment*, 105: 229-259.

Van der Sluijs, J. P., Petersen, A. C., Janssen, P. H. M., Risbey, J. S. and Ravetz, J. R., 2008. "Exploring the quality of evidence for complex and contested policy decisions", *Environmental Research Letters*, 3, 024008 (9pp).

Van der Sluijs, J. P., van Est, R. and Riphagen, M., 2010. "Beyond consensus: reflections from a democratic perspective on the interaction between climate politics and science", *Current Opinion in Environmental Sustainability*, 2(5-6): 409–415.

Van der Sluijs, J.P., Wardekker, J. A., 2015, "Critical appraisal of assumptions in chains of model calculations used to project local climate impacts for adaptation decision support - The case of Baakse Beek", *Environmental Research Letters*, 10(4): 045005.

Zimmer, C., 2015. "Study Finds Climate Change as Threat to 1 in 6 Species", *New York Times*, 30 April.
http://www.nytimes.com/2015/05/05/science/new-estimates-for-extinctions-global-warming-could-cause.html

6

DOUBT HAS BEEN ELIMINATED[1]

Roger Strand

In a speech before the United Nations (UN) Commission on Sustainable Development in 2007, Gro Harlem Brundtland, the UN Secretary-General's Special Envoy on Climate Change, memorably spoke the following words:

> *So what is it that is new today? What is new is that doubt has been eliminated. The report of the International Panel on Climate Change is clear. And so is the Stern report. It is irresponsible, reckless and deeply immoral to question the seriousness of the situation. The time for diagnosis is over. Now it is time to act. (Brundtland, 2007)*

Eight years later, in 2015, barrister and professor of law Philippe Sands proposed that climate sceptics should be dealt with by settling the scientific dispute

---

[1] A previous version of this book chapter was published in S. A. Øyen *et al.* (eds.), *Sacred Science? On science and its interrelations with religious worldviews*, Wageningen Academic Publishers: 55-64. We are grateful to Wageningen Academic Publishers for their kind permission to re-use the material and text.

over climate change in a world court[2]. In the same year *New York Times* columnist Timothy Snyder (2015) expressed the view that climate scepticism was a crime against humanity comparable to the Nazi invasion of Europe and campaign of extermination.

These statements of 2015 were criticized as outrageous, and perhaps rightly so. If they were representative of attitudes in the scientific community, it might seem that climate science was experiencing a moral crisis. However, every climate scientist I have talked to agrees that Nazism was simply and utterly wrong, whereas the climate change issue is complex.

Brundtland's statement is far more interesting because it did *not* provoke a public outcry. Rather, it seemed to emanate from a standpoint shared by many scientific and political actors in the early 21st century: namely a firm belief in 'Science' and great frustration over political inaction on the climate issue. In analysing her statement as well as the curious ethics debate that ensued in Norway—the story of which follows below— we may identify some of the peculiar challenges that arise when science is supposed to speak 'Truth'. I will borrow from Ragnar Fjelland (1985), who employed the Norwegian concept *livssyn* or "life philosophies"[3], and analysed how such life philosophies may borrow au-

---

[2] See *The Guardian*, 2015.
[3] The literal translation of *livssyn* is "life view". One can find three translations of this term into English: (secular) "philosophy of life", "life philosophy" and "worldview". The Norwegian government appears to prefer "philosophy of life" in their official documents. In order to avoid the connotation with the Germanic philosophical tradition, I will use the slightly more awkward "life philosophy" in this chapter. "Worldview" is in my opinion not an adequate translation because the concept of *livssyn* typically embraces and emphasizes the existential and moral dimensions of human life.

thority from science and religion. I shall argue that Brundtland's statement betrays a belief in science that has less to do with philosophy of science and more to do with life philosophy.

Accordingly, in this chapter I shall not be concerned with the supposed 'crisis in science' in the sense of manifestations of moral, epistemological or methodological decline. I will leave it to others to comment on the alleged indecency of thinkers like Sands and Snyder, the loss of numerical skill in scientific culture, and the dubious nature of the mathematical models underlying policy advice—all issues touched on in the present volume. At that level, one may argue either way as to whether science is in crisis or not. On a superficial level, science is clearly not in crisis: scientific research is expanding and scientific knowledge is accumulating—notwithstanding growing concerns about falling quality standards (see Chapter 1 and Chapter 2, this volume). Science is advancing at an accelerating pace and is becoming ever more important in the economy and in public decision-making. There are admittedly examples of public distrust in science, such as the disputes over genetically modified organisms (mainly in Europe and Latin America) or over vaccination programmes (mainly in North America). Still, Northern and Western governments at the beginning of the 21st century, when confronted with resistance to new technologies among their citizenry, do not appear to perceive (or admit to) a crisis in science or modernity. Instead, these governments continue to modernize and rationalize: they dismiss Sands as being ignorant of democracy and freedom of speech; they strengthen controls on research integrity; they fund more and hopefully better science for policy in hope that the resulting knowledge will eliminate political uncertainty. The community of practitioners of post-normal science—already described elsewhere in this volume—would probably argue that such efforts are in vain, and I would agree. However, our arguments have often failed to convince our governments. In this chapter I will analyse one reason for their failure, namely the relation between belief in science and life philosophy. I shall address a type of crisis in science that is metaphysical rather than moral or epistemological.

To posit a metaphysical crisis in science is in no way original. Nietzsche did so in his own manner in the 19th century, and the claim was emphatically and clearly expressed by Husserl in 1936 in his major work *The Crisis of European Sciences*:

> *[…] the total world-view of modern man, in the second half of the nineteenth century, let itself be determined by the positive sciences and be blinded by the 'prosperity' they produced, meant an indifferent turning-away from the questions which are decisive for a genuine humanity. Merely fact-minded sciences make merely fact-minded people. […] What does science have to say about reason or unreason or about us men as subjects of this freedom? The mere science of bodies has nothing to say; it abstracts from everything subjective. (Husserl, 1936/1970: 6)*

For Husserl (and his successors, including Heidegger) the success of Galilean science led not only to a "geometrization" of the life-world in the sense of pervasive measurement and technology, but also to its disenchantment. With modern science, the world became grey. Meaningful places and things became geometric spaces containing bodies and energy fields.

It was perhaps impossible for 19th- and 20th-century philosophers to foresee how new meanings and morals would be created by scientific culture and technological progress. Meaning in the early 21st century is above all connected with welfare and consumption. Policies in the Northern and Western world (as well as in many other parts of the world) are dominated by the idea of research and innovation stimulating economic growth and help-ing to secure long, safe and comfortable lives in welfare states (see Chapter 3). Science produces prosperity, in this view, and also sets the limits for our decisions and actions because it 'speaks Truth to Power'. Husserl rhe-torically asked, "What does science have to say about reason or unreason or about us men as subjects of this freedom?", but in the spirit of Brundtland we might re-ply, "everything". Not only did science provide a meta-physics for modern people, it also provided the basis for a life philosophy.

Is this a crisis? For Husserl and Heidegger the crisis in science was the emptying and to some extent the de-struction or degradation of the life-world. The likes of Brundtland would disagree with this view: they would argue that science has enriched and enhanced the life-world. It is at this point that some philosophical work needs to be done. We need to analyse the way that sci-ence is used in order to inform or create a life philoso-phy. One of the aims of philosophical analysis is to identify contradictions and critically discuss justifica-tions. This is what this chapter sets out to do. I shall dis-cuss Brundtland's claim that doubt has been eliminated, and I intend to show that it represents an ill-founded life philosophy, based on an unscientific faith in science. This philosophy may be added to the "to-be-unlearned" list mentioned in Chapter 1.

The elimination of doubt and the ethos of science

"Doubt has been eliminated," according to Brund-tland, and the reason for this was that the conclusions of the Fourth Assessment Report of the IPCC and "the Stern report" (the "Stern Review on the Economics of Climate Change") were "clear". In other words, these publications were seen (by Brundtland) as carrying sufficient authority to eliminate doubt in the competent and rational reader. It also seems obvious that their authority inhered in their scientific character and credibility. Regardless of how much Brundtland (as a UN Special Envoy, former Prime Minister of Norway and former Director-General of the World Health Organization) may have valued scientific and political authority, the statement that doubt had been eliminated would have been quite ridiculous had it been based merely on, for example, the conclusions of a citizens' consensus conference, the policy of a political party, or a report from an environmentalist NGO. What is noteworthy about the IPCC is that it was set up in order to ensure scientific quality and legitimacy; and the "Stern report" is so called because of Sir Nicholas Stern's status and academic reputation. Their authority does not lie in any notion of their moral superiority but rather in their presumed capacity to *describe things as they are* (or to 'speak Truth to Power'). When Brundtland says that the reports are "clear", it may mean they make clear what needs to be done in terms of decisions and actions, but the clarity she is asserting is first and foremost of a descriptive nature: it *is such that* anthropogenic climate change is upon us.

The problem, however, is that most contemporary philosophies of science—professional philosophies as well as the implicit and informal ones that Kjørup (1996) calls "spontaneous philosophies" — would tend to accord a central place to continuous discussion, open criticism and methodical doubt among their ideals of scientific practice. Indeed, at the very centre of 20th-century expressions of belief in 'Enlightenment' and 'Progress', thinkers such as Karl Popper and Robert K. Merton argued that a critical mind-set and organized scepticism were essential to science and necessary for the maintenance of open and democratic societies. One might certainly disagree with the Popper-Merton theses about the

interdependence of self-critical science and the open society, and dispute the empirical adequacy of describing
scientific institutions and practices as the enactment of
celebrated virtues of open-mindedness. Since the ap-
pearance of the celebrated work of Thomas Kuhn, *The
Structure of Scientific Revolutions* (1962), picking holes in
the theses of Popper and Merton has become an academic industry in itself. Doing so, however, tends to be an
act of distancing oneself from the official discourse on
the 'Ethos of Science' and how it is supposed to be em-
bedded in scientific practice, taught in our universities,
regulated by written and unwritten codes of conduct
and employed in public decision-making. Describing
science as a dogmatic enterprise in the manner of Kuhn
(who talked about "research as a strenuous and devoted
attempt to force nature into the conceptual boxes sup-
plied by professional education", 1962: 5), does not lend
support to the authority of science in public decision
making.

Unsurprisingly, Brundtland was indeed accused, in a
quite literal sense, of an anti-scientific attitude in her
embrace of the IPCC and the Stern reports. In a great
moment of late-modern irony, the Norwegian Research
Ethics Committee for Science and Technology (NENT)
received in November 2009 a complaint about Brundtland's speech which argued that it violated basic principles of research ethics: academic freedom, anti-
dogmatism and organized skepticism[4]. NENT, a committee appointed by the Norwegian government and
mandated by the Norwegian Act on Research Ethics,
concluded in three parts. First, NENT clarified that it

---

[4] The complaint made to NENT by cand. real. ($\approx$ M. Sc.) Jan M.
Döderlein and the Committee's reply are publicly available at
http://www.etikkom.no. I should make clear that I was one of
the 12 members of the NENT Committee at the time and par-
ticipated in the drafting of the response to Döderlein.

would not arbitrate on Brundtland's statement as such because she was not a researcher and the statement apparently did not intend to influence research practice. Second, however, NENT affirmed a relationship between climate research and climate policy. A political speech, regardless of its intentions, may therefore indirectly and unintentionally influence scientific practice. Accordingly, NENT decided that it could and should comment on the content of Brundtland's speech:

> *NENT finds it relevant to point out that accepted language use in scientific contexts differs from what one finds in the mentioned quote [the introductory quote of this chapter]. Traditional academic norms allow and encourage doubt and critical questions. Doubt may in such contexts be well or ill founded, but not irresponsible and immoral by itself. In a situation of action, which then is not a purely scientific context, it may of course be irresponsible and immoral not to act, for example by maintaining doubt or criticism that one oneself finds poorly justified. It might be that Brundtland has this type of action in mind. (NENT, 2009)[5]*

The statement from NENT goes on to mention precautionary principles as an example of principles of action designed to alleviate the tension between academic and political moral norms: "Such principles seek to justify political action while acknowledging scientific uncertainty and maintaining critical scientific debate" (*ibid.*).

In sum, NENT probably went as far as it could within its mandate, concluding that Brundtland's speech differed from "accepted language use in scientific contexts". In plain terms: her utterance violated the norms of the ethos of science. It would be a serious underestimation of actors at Brundtland's level, however, to think that her words were carelessly chosen or the

---

[5] Author's own translation.

result of ignorance. She certainly knew that it is more 'scientific' to qualify statements; to appreciate the plurality of perspectives and expert opinions; to show awareness of the essential fallibility of scientific facts, theories and advice. Her task was not to be scientific: it was to argue for the supreme authority of science in order to quash doubts about the authority of the advice from the IPCC and the Stern report. Whether this was a wise or even effective strategy, is another question (Sarewitz, 2004).

The unscientific belief in science

Questions about the justification of the authority of various perspectives and positions are very difficult to answer and have received extensive attention from (some would say *have plagued*) modern philosophy. I shall not enter here into what many would claim to be the more fundamental issue — namely how to justify one's own special beliefs (what some would call *comprehensive doctrines*) in a politically liberal society where others cannot be expected to share one's world-view or to endorse the same set of values or virtues. Rather, I will discuss some aspects of justification *from within* a particular perspective in the hope that it will shed some light on the topic of this chapter, that is, the relationship between science and life philosophies as exemplified by Brundtland's speech.

The task of justifying a comprehensive doctrine from within that doctrine can range from the trivial to the extremely difficult. For instance, doctrines that postulate their own origins in revelations made by an omniscient, loving and truthful deity can have strong self-justificatory features: it is natural to believe God's words if they tell us that he is always right. Proponents of doctrines about the proper role and authority of science can choose from a number of justificatory strategies. Justificatory resources can sometimes be found within the perspective itself, as in the intriguing debate on the evidence for the utility of evidence-based practice in medicine. At other times, it has been found convenient to emphasize that justification in the last resort resides outside the perspective, as when Popper pointed out the need to *decide* upon the role of rationality and the choice of critical rationalism. Critical rationalism is not consistent with claiming the *necessity* of its acceptance, if we are to believe Popper.

This is a relevant observation in relation to Brundtland's speech. It is possible to have a scientific belief in science — but if science is defined epistemologically as fallible and praxeologically as an activity that embodies norms of doubt and self-criticism, then belief in science can be neither too dogmatic nor too hostile towards criticism without becoming unscientific. This paradox is indeed evident in Brundtland's statement. She claims not only that "Doubt has been eliminated" but also that to raise further critical questions would be immoral. It is very difficult not to see this as expressly unscientific and even at odds with the norms of the institutions from which she borrows authority for her statement. The contradiction is perhaps not so important in itself. There is little reason to fear that climate scientists will become dogmatic simply because one UN Special Envoy made an unscientific claim about climate science. The interesting question is rather: if science is not the source of authority for this type of belief in science, what is? Mere trust in the IPCC or in Stern and his team, however brilliant they may be, appears an inadequate basis for making such strong claims. Many observers would find it hard to trust such a complex and earthly endeavour as the IPCC to the point of not admitting the slightest doubt. Bearing in mind Brundtland's experience as former Head of Government and former Director-General of the WHO, it is unlikely that she held naïve beliefs about the functioning of large international institutions. On this ground, the statement "Doubt has been eliminated" appears less an expression of reasonable trust in the mundane IPCC than of faith in science. What kind of phenomenon was that faith?

Livssyn — life philosophies

There is an abundance of potentially useful concepts for the problems I am discussing here. *Comprehensive doctrine* is one example. *Ideology, metaphysical position* and *worldview* are others. The point I wish to pursue, however, is not so much one of epistemology or political theory as of "life philosophy" in Fjelland's definition. In the following, I shall discuss his analysis, as well as the Norwegian context in which it was introduced.

At the time of Brundtland's speech in 2012, Norway was a confessional state. The Norwegian Constitution read:

> *All inhabitants of the Realm shall have the right to free exercise of their religion. The Evangelical-Lutheran religion shall remain the official religion of the State. The inhabitants professing it are bound to bring up their children in the same. (Norwegian Constitution (as of 2012), Article 2)*

The wording of the constitution was changed in 2014; it now reads that "our value foundation remains our Christian and humanistic heritage". The "State Church" changed its name to the "Church of Norway" and is mentioned in the Constitution as the "people's church". In practice, however, the State can still be regarded as confessional, considering that clergy are civil servants, Christianity is taught in public schools, *etc*. This is not without practical complications in a modern welfare state. For instance, the State needs to distinguish between inhabitants who profess the official religion and those who do not, in order to keep track of public church taxes. A proportion of direct taxes collected from Church of Norway members is re-directed to the church, whereas other, non-State religious communities receive a share of public taxes proportionate to the size of their congregations. This principle is intended to uphold the constitutional right to free exercise of religion. The economic and institutional dimension of the "exercise of religion" is accordingly governed by membership in religious communities, provided that the community is entered in the designated official registry of such communities. Entry in the registry (and hence, the right to receive part of the church tax) has to be approved by the State, and is regulated by the Norwegian Act with the striking title *Lov om trudomssamfunn og ymist anna* ("Norwegian Act on Religious Communities, *et cetera*" [*sic*]). In 1981, secular communities were mentioned by name in Norwegian legislation (and not only as "*et cetera*") in the "Act Relating to Allocations to Belief-based Communities", and in public management the relevant category is now more often than not *tros- og livssynssamfunn*: that is, faith/religious and (secular) belief/life philosophy-based communities. A lot could be said about this, not least about the curious controversies that occasionally arise when the State decides not to approve a particular community as fit to be listed in the official state registry. For instance, the County Governor of Telemark withdrew the approval of "The Circle of Friends of Pi-ism" in 2006, referring to media reports that the community could not be not a *serious* religious community[6] as they were "laughing" about their own approved status.

In the obviously difficult work of deciding what constitutes a serious religious community, the Norwegian

---

[6] See http://www.fylkesmannen.no/liste.aspx?m=5783&amid =1303323

State relies on a definition provided by its Ministry of Justice. The definition is quite comprehensive and requires that the (true) followers of a (serious) religion believe in a power or powers that determine fate, and that they lead their lives accordingly. Moreover, a religion "should" include fundamental concepts such as "holiness", "revelation", "miracle", "sin" and "sacrifice" (Kirke-, utdannings og forskningsdepartementet, 2000; see p. 50 for a self-critical discussion of this position). As for secular communities, it appears that Norwegian authorities never even tried to define them, opting instead for the more pragmatic solution of setting a lower limit of 500 members as a requirement for their approval. Still, the concept is not empty and appears to be endowed with an implicit expectation of seriousness and dignity, as indicated by Norway's Prime Minister Jens Stoltenberg in a speech to Parliament on 28 May 2010:

President,

> *Religion and life philosophy have always been an important part in the life of human beings. Our relationship to religion and life philosophy takes part in defining us both as individuals and as a society. [He then proceeds to describe religious life.] [...] Others do not believe in a deity, but find a sense of belonging and guidance in a distinct life philosophy. (Stoltenberg, 2010)[7]*

Hence, the sincere follower of a life philosophy — humanism and social humanism being the most visible ones in Norwegian public life — should perhaps not necessarily profess *faith,* but at least "belonging" and "guid-

---

[7] Author's translation, which was anything but easy in this case. Stoltenberg uses the word *tydelig — deutlich* in German. I have translated it into "distinct", which perhaps exaggerates the association with rationalist philosophy in his mention of secular life philosophies.

ance", and should certainly not make a joke of his or her life philosophy or community.

However, even a close reading of Norwegian public documents is not very enlightening about what a life philosophy actually is. The impression that emerges is that a life philosophy is defined by what it *is not*: life philosophy is like a religion, but without religious beliefs; 'Religion Light' or even 'Zero'. At this point, Fjelland's (1985) analysis comes to the rescue.

Fjelland argues that Kant's four questions of philosophy are the suitable point of departure for defining a life philosophy:

  i. What can I know?
  ii. What ought I to do?
  iii. For what may I hope?
  iv. What is a human being?

Rather than reproduce Fjelland's argument, I shall apply his conclusions in the final part of the chapter, namely that one's individual answers to the three latter questions form a life philosophy. The answer to the first question — what can I know? — does not form an intrinsic part of the life philosophy, but may be central to its justification.

In this way, the concept of life philosophy is placed on a different level to religion and science. Religion and science may *provide* inputs to (or justifications for) the life philosophy, but they are not identical to the life philosophy. Fjelland shows how not only a religion such as Christianity but also a cosmology such as is found in Ancient Greek philosophy can provide answers to Kant's questions, and in this way justify particular life philosophies (from within the perspective itself, of course). He then argues that belief in science and progress can easily provide other answers to Kant's questions and in this way produce a science-based life philosophy. 'Science-based' is, however, a term to be used with caution in this respect. *Within* the proper domain of science, the quality of being 'science-based' may endow a claim with superior epistemic authority. But Kant's questions are philosophical, not scientific; a categorical mistake is made if one believes that biological theories can produce the unique and ultimate truth about human beings, or if one uncritically embraces the technological imperative, to conclude that we ought to develop and implement all technologies that can be delivered by science.

First or second modernity

Is it appropriate to talk about life philosophies in relation to climate change? I think so. The issue of climate change cannot be separated from a number of immense and difficult questions about our responsibility for future generations, global equity, non-human species, our choice of lifestyles and therefore our values. I believe Brundtland would agree on this point. She was not pretending to be a philosopher of science. She wanted to make a statement about what was important and how we ought to act as societies and individuals.

Brundtland's speech borrows the answers to Kant's second and third questions — what we ought to do and for what we may hope — from the IPCC and the Stern reports: we should reduce emissions, for the beneficial effects. She touched on the question of what a human being is in the report *Our Common Future* (WCED, 1987), which not only lays out our responsibilities regarding future generations and present neighbours, but also sees our roles and identities as intrinsically bound together on Planet Earth. Many would agree with her.

A problem arises when considering Kant's first question: what can we know? By expressing an unscientific faith in science, Brundtland undermines the authority of her life philosophy. It remains science-based, but no longer justified and endorsed by science in its canonical expression. Nor is it supported by religion. Perhaps it could be supported by a 'cosmology of simplicity', in which the Universe is, if not a Greek Harmonic Cosmos, at least such a simple place that even if scientists remain mired in methodological self-doubt, the knowledge they produce is in fact the Truth. This position is an intellectually vulnerable one, to the left of Religion and to the right of Science. Popper tried to find a way out if it; Feyerabend spent most of his intellectual energy undressing and ridiculing it. The problem for Brundtland and other policy makers is not the keen eye of philosophers of science; the problem is that many 21st-century citizens are endowed with critical skills and literacy and fear authority so little that they no longer believe leaders such as Brundtland when they say that doubt has been eliminated and that doubt is, moreover, immoral. Interpreted as an empirical statement, the pronouncement "doubt has been eliminated" is quite simply false. The maker of such a statement can only be held up for ridicule, and his or her communicative power will suffer. The serious diminution of political agency in the climate issue has been analysed elsewhere (Rommetveit, Funtowicz and Strand, 2010) and I shall not enter into political analysis here. I shall only recall Ulrich Beck's more general diagnosis of the class of political-environmental-human global problems that increasingly appear to characterize our century:

> *Wir leben in einer anderen Welt als in der, in der wir denken. Wir leben in der Welt des und, denken in Kategorien des entweder-oder. [...] Die stinknormale Weiter–und–weiter-Modernisierung hat einen Kluft zwischen Begriff und Wirklichkeit aufgerissen, die deshalb so schwer aufzuzeigen, zu benennen ist, weil die Uhren in den zentralen Begriffen stillgestellt sind. Die "Moderne"*

[...] ist in ihrem fortgeschrittenen Stadium zur terra incognita geworden, zu einer zivilisatorischen Wildnis, die wir kennen und nicht kennen, nicht begreifen können, weil das monopolistischen Denkmodell der Moderne, ihr Industriegesellschaftliches, industriekapitalistisches Selbstverständnis, im Zuge der verselbstständigten Modernisierung hoffnungslos veraltet ist. (Beck, 1993: 61-62)[8]

The Norwegian Research Ethics Committee for Science and Technology mentioned earlier has called for a 'second modernity' type of approach (NENT, 2009): that is, admitting that there may be uncertainty in the climate models but affirming that this does not justify inaction; indeed, that uncertainty may be a *reason* for precautionary action. Brundtland's problem is that she does not find enough power in a discourse of 'ands': Science is telling us that the climate problem is extremely urgent *and* that science may be wrong or incomplete. Apparently unable to acknowledge the second modernity, and no longer able to 'educate and persuade' the people into submissiveness, leaders of 21st-century democracies are simply in deadlock.

---

[8] This passage is not found in the official English translation of Beck's book, probably because it does not translate at all well into English. What follows is an attempt of the author and Sarah Moore, copy-editor of this volume:

We live in a different world than in the one in which we think. We live in the world of 'ands', while thinking in terms of 'either-or'. [...] The perfectly banal process of modernization goes on and on and has torn open a chasm between ideas and reality which is so hard to point out, so hard to express, because the clocks have stopped in the central ideas. The 'modern' [...] in its advanced stages has become a terra incognita, it has become a civilizational wilderness that we know and do not know, that we cannot understand because the monopolistic conceptual model of modernity, its industrialized-society, capitalistic self-understanding, has become hopelessly outdated even as the process of modernization has established itself.

How can we get out of this deadlock? There have already been many constructive suggestions, of various natures: methodological (responsible quantification and appraisal of uncertainties/sensitivities), epistemological (management of uncertainty and complexity), political (deliberative democracy and a new social contract for science), legal (precautionary principles), ethical (eco-philosophy, *etc.*). This short chapter shall end with a somewhat unusual suggestion: to think again about our concepts of life philosophies. We have seen that the Norwegian Prime Minister could not avoid quasi-religious concepts such as 'belonging' and 'guidance' when trying to describe life philosophies. As long as this deferential attitude prevails, secular life philosophies will remain a version of Religion Light. We will be prone to fall into science-based but unscientific dogmatism and consequently into ridicule. I will therefore end by making the claim for piecemeal, reflexive, self-critical and tentative life philosophies, accommodating the 'ands' of Beck and making room for doubts and smiles. A proper argumentation would require another chapter, or indeed a book series; still, let me put forward the claim that a life philosophy of Beck's 'ands', fit for a second modernity, would need to be able to maintain hope in the absence of guarantees from gods or science, and to see the questions of how we should act and what it is to be a human as deeply entangled and interdependent (Funtowicz and Strand, 2011). Such a philosophy would need to be founded in the 21st-century life-world, accepting that science and technology generate meaning and purpose but without accepting the hegemonic technoscientific imperative. Let me put forward this claim for a new life philosophy, even if a community of such thinkers might not be found fit by the County Governor of Telemark, Norway, to enter the appropriate State registry.

References

Beck, U., 1993. *Die Erfindung des Politischen: zu einer Theorie reflexiver Modernisierung*. Frankfurt am Main: Suhrkamp.

Brundtland, G. H., 2007. Speech at the UN Commission on Sustainable Development. http://www.regjeringen.no/en/dep/ud/selected-topics/un/Brundtland_speech_CSD.html?id=465906.

Fjelland, R., 1985. *Vitenskap som livssyn? Livssynsspørsmål i teknikkens tidsalder*. Tromsø: Universitetsforlaget.

Funtowicz, S. and Strand, R., 2011. "Change and Commitment: Beyond Risk and Responsibility", *Journal of Risk Research*, 14: 995-1003.

Guardian, 2015. "World court should rule on climate science to quash sceptics, says Philippe Sands", 18 September. http://www.theguardian.com/environment/2015/sep/18/world-court-should-rule-on-climate-science-quash-sceptics-philippe-sands?CMP=twt_a-science_b-gdnscience.

Husserl, E., 1970 (1936). *The crisis of European sciences and transcendental phenomenology*. Evanston, IL: Northwestern University Press.

Kirke-, utdannings- og forskningsdepartementet [Ministry of Church, Education and Research] 2000. "Tro og livssyn som grunnlag for tilskudd og offentlige funksjoner. Rapport" ["Faith and life philosophy as a basis for public financial support and public functions. Report"]. Oslo: Kirke-, utdannings- og forskningsdepartementet. http://www.regjeringen.no/upload/KD/Vedlegg/Rapporter/troog050.pdf

Kjørup, S., 1996. *Menneskevidenskaberne: problemer og traditioner i humanioras videnskabsteori*. Frederiksberg: Roskilde Universitetsforlag.

Kuhn, T. 2012 (1962). *The Structure of Scientific Revolutions*. Chicago, IL: University of Chicago Press.

NENT, 2009. "Bør handle tross vitenskapelig usikkerhet" ["Should act in spite of scientific uncertainty"]. http://www.etikkom.no/no/Aktuelt/Aktuelt/Nyheter/2009/Bor-handle-tross-vitenskapelig-usikkerhet/

Rommetveit, K., Funtowicz, S. and Strand, R., 2010. "Knowledge, democracy and action in response to climate change", in: Bhaskar, R., Frank, C., Høyer, K. G., Naess, P.

and Parker, J., *Interdisciplinarity and Climate Change*: 149-163. Abingdon: Routledge.

Sarewitz, D., 2004. "How science makes environmental controversies worse". *Environmental Science & Policy* 7: 385-403.

Snyder, T., 2015. "The Next Genocide", *New York Times*, 12 September.
http://www.nytimes.com/2015/09/13/opinion/sunday/the-next-genocide.html

Stoltenberg, J., 2010. "Religionens plass i samfunnet" ["The Place of Religion in Society"], speech to the Norwegian Parliament 28 May 2010.
http://www.regjeringen.no/nb/dep/smk/aktuelt/taler_og_artikler/statsministeren/statsminister_jens_stoltenberg/2010/Religionens-plass-i-samfunnet.html?id=606053

WCED (World Commission on Environment and Development), 1987. *Our Common Future*. Oxford: Oxford University Press.

ABOUT THE AUTHORS

Alice Benessia holds a Ph.D. in Science, Technology and Law, an M.A. in the Philosophical Foundations of Physics and an M.F.A in Photography and Related Media. She is a Research Fellow in the Epistemology of Sustainability at the Interdisciplinary Research Institute on Sustainability (IRIS) at the University of Torino. She has been an appointed expert at the Joint Research Centre of the European Commission. She is a founding member of the Italian Association for Sustainability Science. She also works as a visual artist and has taught and lectured on photography and visual arts in numerous participatory workshops. Her interdisciplinary research deals with epistemological issues arising in the framework of art, science and sustainability, with special interest in visual language.

Silvio Funtowicz taught mathematics, logic and research methodology in Buenos Aires, Argentina. During the 1980s he was a Research Fellow at the University of Leeds, England. Until his retirement in 2011 he was a scientific officer at the Institute for the Protection and Security of the Citizen (IPSC) of the Joint Research Centre of the

European Commission (EC-JRC). Since February 2012 he has been Professor II at the Centre for the Study of the Sciences and the Humanities (SVT) at the University of Bergen, Norway. He is the author of numerous books and papers in the field of environmental and technological risks and policy-related research. He has lectured extensively and is a member of the editorial board of several publications and the scientific committee of many projects and international conferences.

Mario Giampietro is ICREA (Catalan Institution for Research and Advanced Studies) Research Professor at the Institute of Environmental Science and Technology (ICTA) of the Autonomous University of Barcelona (UAB), Spain. He works on integrated assessment of sustainability issues using new concepts developed in complex systems theory. He has developed an innovative scientific approach called Multi-Scale Integrated Analysis of Societal and Ecosystem Metabolism (MuSIASEM) integrating biophysical and socio-economic variables across multiple scales, thus establishing a link between the metabolism of socio-ecological systems and the potential constraints of the surrounding natural environment. Recent research has focused on the nexus between land use, food, energy and water in relation to sustainable development goals. He has authored or co-authored over two hundred publications, including six books, on research themes such as multi-criteria analysis of sustainability; integrated assessment of scenarios and technological changes; alternative energy technologies; energetics; urban metabolism; biocomplexity and sustainability; and science for governance.

210

Ângela Guimarães Pereira holds a Ph.D. in Environmental Systems and their Tensions. She began working at the Joint Research Centre of the European Commission in 1996 on European projects focussing on environmental and societal issues, future-oriented activities  and the integration of information technologies with public engagement. Her work has been inspired by the post-normal science framework developed by Silvio Funtowicz and Jerome Ravetz in the 1990s. At the JRC she currently works on knowledge assessment and the ethics of technoscience, critically investigating their governance and their correspondence with current innovation narratives. She has authored many peer-reviewed publications and was co-editor of *Interfaces between Science & Society* with Greenleaf in 2006, *Science for Policy: Challenges and Opportunities* with Oxford University Press in 2009 and *The End of the Cartesian Dream* with Routledge in 2015. Her current interests lie in the history of science and the emergence and reinvention of ways of knowing. Her favourite story is Hans Christian Andersen's "The Emperor's New Clothes".

Jerome R. Ravetz is a leading authority on the social and methodological problems of contemporary science. He was born in Philadelphia and attended Central High School and Swarthmore College. He came to England on a Fulbright scholarship, where he did a  Ph.D. in Pure Mathematics at Trinity College, Cambridge. He lectured in the History and Philosophy of Science at Leeds University. He is currently an Associate Fellow at the Institute for Science, Innovation and Society at Oxford University. With Silvio Funtowicz he created the NUSAP notational system for assessing the uncertainty and quality of scientific information, in *Uncertainty and Quality in Science for Policy*, and also the concept of post-normal science, relevant when "facts are uncertain, values in dispute, stakes high and decisions urgent". His earlier seminal work, *Scientific Knowledge and its Social Problems* (Oxford U.P., 1971 and Transaction, 1996), now has a shorter sequel, *The No-Nonsense Guide to Science* (New Internationalist, 2006). His other publications include a collection of essays, *The Merger of Knowledge with Power* (Mansell 1990). He is currently working on a 'New Arithmetical Language for Policy', based on 'soft numbers', employing sparse arithmetic and dynamical graphical methods.

Andrea Saltelli has worked in physical chemistry, environmental sciences, applied statistics, impact assessment and science for policy. His main disciplinary focus is on sensitivity analysis of model output, a discipline in which statistical tools are used to interpret the output  of mathematical or computational models, and on sensitivity auditing, an extension of sensitivity analysis to the entire evidence-generating process in a policy context. A second focus is on the construction of composite indicators or indices. Until February 2015 he led the Econometric and Applied Statistics Unit of the European Commission at the Joint Research Centre in Ispra, Italy, developing econometric and statistical applications, mostly in support to the services of the European Commission, in fields such as lifelong learning, inequality, employment, competitiveness and innovation. He has been involved in training European Commission staff on impact assessment. At present he is at the European Centre for Governance in Complexity (ECGC), a joint undertaking of the Centre for the Study of the Sciences and the Humanities (SVT) at the University of Bergen (UIB) and the Institute of Environmental Science and Technology (ICTA) at the Autonomous University of Barcelona (UAB). The ECGC is located in the UAB campus in Barcelona.

Daniel Sarewitz is Professor of Science and Society and co-director and co-founder of the Consortium for Science, Policy & Outcomes (CSPO) at Arizona State University. He is interested in the relationships among knowledge, technology, uncertainty, disagreement, policy and social outcomes. His most recent book is *The Techno-Human Condition* (co-authored with Braden Allenby, MIT Press, 2011). He is the editor of the magazine *Issues in Science and Technology* (www.issues.org), and also a regular columnist for *Nature*. From 1989-1993 he worked on R&D policy issues for the U.S. House of Representatives Com-mittee on Science, Space, and Technology. Together with the writer Lee Gutkind, he is currently starting up a new project on science and religion.

Roger Strand originally trained as a natural scientist (with a Ph.D. in biochemistry, 1998), then developed research interests in the philosophy of science and issues of scientific uncertainty and complexity. This has gradually led his research into broader areas of social research and philosophy, including questions of policy, decision-making and governance at the science-society interface.

Strand is currently Professor at the Centre for the Study of the Sciences and Humanities at the University of Bergen, Norway. He has coordinated two EU FP7 projects (TECHNOLIFE and EPINET) that addressed the need for more dynamic governance of science in society. He was a member of the National Research Ethics Committee for Science and Technology in Norway (2006-13) and Chair of the European Commission Expert Group on Indicators for Responsible Research and Innovation (RRI; 2014-15).

Jeroen P. van der Sluijs is Professor in the Theory of Science and Ethics of the Natural Sciences at the University of Bergen and Associate Professor in new and emerging risks at Utrecht University. His research focusses on scientific controversy around environmental and health risks in situations where scientific assessment is used as a basis for policy making before conclusive scientific evidence is available on the causal relationships, the magnitude, and the probabilities of these risks. His work seeks to understand and improve the science–policy interface in a context of deep uncertainty by contributing and applying deliberative methods and tools for the assessment of the quality of knowledge. He has been working on contested science in the fields of climate change, pollinator decline, fish stock assessments, endocrine disruptors, electromagnetic fields, nanoparticles, underground storage of CO_2 and risk migration in sustainable technologies. Jeroen has published 78 articles in peer-reviewed scientific journals and 27 peer-reviewed book chapters.

Affiliations

Alice Benessia (1)

Silvio Funtowicz (2)

Mario Giampietro (3, 4, 5)

Ângela Guimarães Pereira (6)

Andrea Saltelli (2, 4, 5)

Jerome Ravetz (7)

Daniel Sarewitz (8)

Roger Strand (2, 5)

Jeroen P. van der Sluijs (2)

---

(1) Interdisciplinary Research Centre on Sustainability (IRIS), University of Torino

(2) Centre for the Study of the Sciences and the Humanities (SVT), University of Bergen (UIB)

(3) Institució Catalana de Recerca i Estudis Avançats / Catalan Institution for Research and Advanced Studies (ICREA)

(4) Institut de Ciència i Tecnologia Ambientals / Institute of Environmental Science and Technology (ICTA), Universitat Autonoma de Barcelona (UAB)

(5) European Centre for Governance in Complexity, UAB Bellaterra Campus

(6) European Commission–Joint Research Centre, Ispra

(7) Institute for Science, Innovation and Society, University of Oxford

(8) Consortium for Science, Policy & Outcomes at Arizona State University

Made in the USA
Lexington, KY
17 April 2016